RITA *Intelligent Transportation Systems Joint Program Office*

The Design of a Minimal Sensor Configuration for a Cooperative Intersection Collision Avoidance System – Stop Sign Assist:

(CICAS-SSA Final Report # 4)

August 2010

U.S. Department of Transportation
Research and Innovative Technology Administration

Intelligent Vehicles Laboratory
Department of Mechanical Engineering
University of Minnesota
111 Church Street SE
Minneapolis, MN 55455

Prepared by

Alec Gorjestani
Arvind Menon
Pi-Ming Cheng
Craig Shankwitz
Max Donath

Notice

This document is disseminated under the sponsorship of the U.S. Department of Transportation in the interest of information exchange. The U.S. Government assumes no liability for the use of the information contained in this document. This report does not constitute a standard, specification, or regulation.

The U.S. Government does not endorse products or manufacturers. Trademarks or manufacturers' names appear in this report only because they are considered essential to the objective of the document.

The contents of this report reflect the views of the authors, who are responsible for the facts and the accuracy of the information presented herein. This document is disseminated under the sponsorship of the Department of Transportation University Transportation Centers Program, in the interest of information exchange. The U.S. Government assumes no liability for the contents or use thereof. This report does not necessarily reflect the official views or policies of the Minnesota Department of Transportation, the Intelligent Transportation Systems Institute, or the University of Minnesota.

1. Report No.	2. Government Accession No.	3. Recipient's Catalog No.

4. Title and Subtitle	5. Report Date
The Design of a Minimal Sensor Configuration for a Cooperative Intersection Collision Avoidance System – Stop Sign Assist: CICAS-SSA Report #2	August 2010
	6. Performing Organization Code:

7. Author(s)	8. Performing Organization Report No.
Alec Gorjestani, Arvind Menon, Pi-Ming Cheng, Craig Shankwitz, Max Donath	CTS Project #2006050

9. Performing Organization Name and Address	10. Work Unit No.
Department of Mechanical Engineering University of Minnesota 111 Church Street S.E. Minneapolis, Minnesota 55455	
	11. Contract or Grant No.

12. Sponsoring Agency Name and Address	13. Type of Report and Period Covered
United States Department of Transportation, Federal Highway Administration 1200 New Jersey Ave, S.E. Washington, DC 20590	Work started March 2008. Draft report submitted to Mn/DOT: January 2009
	14. Sponsoring Agency Code

15. Supplementary Notes

16. Abstract

The deployment of a Cooperative Intersection Collision Avoidance System – Stop Sign Assist (CICAS-SSA) can save lives by addressing the causal factor of crashes at rural thru-Stop intersection: drivers who stop on the minor leg of the intersection, improperly assess the gaps in the traffic on the major leg, proceed, and are then hit.

The prototype CICAS-SSA system consisted of a network of sensors covering both the minor and the major legs of the intersection. Sensors on the minor road monitored the approach of vehicles and classified them based on their length and height. Sensors along the major road were arrayed to track vehicles (and the gaps between them) approaching the crossroads from 2000 feet away as a means to ensure that the tracking algorithm had sufficient time to "lock on" and track all approaching vehicles.

Because cost is a primary concern for any highway safety application, the development of a "minimal sensor set" which would provide adequate safety performance for minimum cost was paramount to the success of the CICAS-SSA program. This report documents the development of this minimal sensor configuration.

17. Key Words	18. Distribution Statement
Rural highways, Unsignalized intersections, Highway safety, Collisions, Gap acceptance, Cooperative Intersection Collision Avoidance Systems-Stop Sign Assist (CICAS-SSA) program, Warning signs, Sensors, Traffic surveillance, Arterial highways, Secondary roads	No restrictions. Document available from: National Technical Information Services, Springfield, Virginia 22161

19. Security Class if. (of this report)	20. Security Class if. (of this page)	21. No. of Pages	22. Price
Unclassified	Unclassified	49	

Form DOT F 1700.7 (8-72)   Reproduction of completed page authorized

TECHNICAL REPORT DOCUMENTATION PAGE

Acknowledgements

This work is funded by the United States Department of Transportation Federal Highway Administration (US DOT FHWA) and the Minnesota Department of Transportation (Mn/DOT) through Cooperative Agreement DTFH61-07-H-00003, and by State Pooled Fund Project TPF-5(086).

Listed below are the currently available reports in the CICAS-SSA Report Series (as of August 2010):

Determination of the Alert and Warning Timing for the Cooperative Intersection Collision Avoidance System – Stop Sign Assist Using Macroscopic and Microscopic Data: CICAS-SSA Report #1
Prepared by: Alec Gorjestani, Arvind Menon, Pi-Ming Cheng, Craig Shankwitz, and Max Donath

The Design of a Minimal Sensor Configuration for a Cooperative Intersection Collision Avoidance System – Stop Sign Assist: CICAS-SSA Report #2
Prepared by: Alec Gorjestani, Arvind Menon, Pi-Ming Cheng, Craig Shankwitz, and Max Donath

Macroscopic Review of Driver Gap Acceptance and Rejection Behavior at Rural Thru-Stop Intersections in the U.S. – Data Collection Results in Eight States: CICAS-SSA Report #3
Prepared by: Alec Gorjestani, Arvind Menon, Pi-Ming Cheng, Bryan Newstrom, Craig Shankwitz, and Max Donath

Sign Comprehension, Considering Rotation and Location, Using Random Gap Simulation for Cooperative Intersection Collision Avoidance System – Stop Sign Assist: CICAS-SSA Report #4
Prepared by: Janet Creaser, Michael Manser, Michael Rakauskas, and Max Donath

Validation Study – On-Road Evaluation of the Cooperative Intersection Collision Avoidance System – Stop Sign Assist Sign: CICAS-SSA Report #5
Prepared by: Michael Rakauskas, Janet Creaser, Michael Manser, Justin Graving, and Max Donath

Additional reports will be added as they become available.

TABLE OF CONTENTS

- **1 Introduction** 1
- **2 Comprehension StudIES** 4
  - 2.1 Experiment One Methods 6
    - 2.1.1 Participants 6
    - 2.1.2 Materials 8
    - 2.1.3 Procedures 8
    - 2.1.4 Statistics 10
  - 2.2 Experiment One Results 11
    - 2.2.1 Inter-rater Reliability 11
    - 2.2.2 Comprehension Results 13
  - 2.3 Experiment One Conclusions 29
    - 2.3.1 Countdown Messages 29
    - 2.3.2 Icon Sign 30
    - 2.3.3 Hazard Sign 31
  - 2.4 Experiment Two Methods 31
    - 2.4.1 Participants 31
    - 2.4.2 Apparatus 32
    - 2.4.3 Procedures 32
    - 2.4.4 Statistics 34
    - 2.4.5 Results 34
    - 2.4.6 Countdown Sign 34
    - 2.4.7 Icon Signs 37
    - 2.4.8 Hazard Sign 39
  - 2.5 Experiment Two Conclusions 40
    - 2.5.1 Countdown Sign 40
    - 2.5.2 Icon Sign 41
    - 2.5.3 Hazard Sign 41
  - 2.6 Discussion 41
    - 2.6.1 Countdown Sign 42
    - 2.6.2 Icon Sign 43
    - 2.6.3 Hazard Sign 43
    - 2.6.4 Age 44
  - 2.7 Conclusions 44
  - 2.8 Recommendations 45
    - 2.8.1 Countdown Sign 45
    - 2.8.2 Icon Sign 46
    - 2.8.3 Hazard Sign 47
- **3 Rotation Study** 48
  - 3.1 Methods 49
    - 3.1.1 Participants 49
    - 3.1.2 Apparatus 49
    - 3.1.3 Driving Scene 50
    - 3.1.4 Procedures 50
    - 3.1.5 Analyses 51
  - 3.2 Results 51
    - 3.2.1 Comprehension 51
      - 3.2.1.1 Accuracy in Mapping SSA Sign Information to Traffic Conditions 51
      - 3.2.1.2 Confidence in Identifying Traffic that the Sign is Informing About 52
    - 3.2.2 Usability 53
      - 3.2.2.1 Easy to Associate Information on the Sign to Traffic Conditions 53
      - 3.2.2.2 Comfortable to View Sign in this Location 53

      3.2.2.3   Obstructed View Approaching Traffic 54

      3.2.2.4   Viewing Angle that Best Maps to Roadway 55

  3.3   Discussion 57

  3.4   Conclusions 58

4 Location Study 59

  4.1   Methods 60

    4.1.1   Participants 60

    4.1.2   Procedures 61

    4.1.3   Analyses 61

  4.2   Results 62

    4.2.1   Comprehension 62

      4.2.1.1   Timed Comprehension Response Behavior (Response Time; RT) 62

      4.2.1.2   Accuracy - Ease of Mapping Information to Traffic Conditions 62

      4.2.1.3   Confidence in Identifying Traffic that the Sign ss Informing About 64

    4.2.2   Comprehension Summary 65

    4.2.3   Usability 65

      4.2.3.1   It Was Easy to Associate Information on the Sign to Traffic Conditions 66

      4.2.3.2   It Was Comfortable to View Sign in This Location 66

      4.2.3.3   The Sign Obstructed My View of Approaching Traffic 66

      4.2.3.4   It Was Easy to See at this Distance 66

      4.2.3.5   Usability Summary 67

    4.2.4   Comparison Preferences 67

      4.2.4.1   Layout Set Preference 67

      4.2.4.2   Sign Location Preference While Waiting at the Stop Sign 68

      4.2.4.3   Sign Location Preference While Waiting in the Median 70

      4.2.4.4   Comparison Preferences Summary 70

  4.3   Discussion 71

  4.4   Conclusions 72

5 Random Gap Study 74

  5.1   Methods 74

    5.1.1   Participants 74

    5.1.2   SSA Interfaces 75

    5.1.3   Driving Simulator 75

    5.1.4   Simulated Test Intersection 75

    5.1.5   Warning Thresholds 75

    5.1.6   Randomized Traffic Streams 76

    5.1.7   Procedures 77

    5.1.8   Statistics 79

  5.2   Results - Driving Performance 81

    5.2.1   Accepted Gaps 81

    5.2.2   Rejected Gaps 82

    5.2.3   Safety Margins 82

    5.2.4   Movement Time 84

    5.2.5   Wait Time 84

    5.2.6   Crossing Maneuver Type 85

    5.2.7   Crashes86

  5.3   Results - Subjective 86

	5.3.1	Post-Drive Questionnaire	86
	5.3.2	Usability Scales	87
	5.3.3	Sign Use	88
	5.3.4	Sign Preference	88
	5.4	Discussion	88
	5.4.1	Age	91
	5.4.2	Light Conditions	91
	5.5	Limitations	92
	5.6	Conclusions	92

6 Overall Conclusions 93
    6.1   Benefits of Validation in a Simulated Driving Context 94

References 96

LIST OF FIGURES

Figure 1. Diagram of a stop-controlled trunk-highway thru-stop intersection with relative location of SSA signs. 1

Figure 2. Diagram used in Study Introduction to show the rural, stop-controlled intersection.... 10

Figure 3. Depiction of the Do Not Enter message options. 13

Figure 4. Depiction of the Do Not Cross/Turn Left message options. 13

Figure 5. Depiction of the Proceed with Caution message options. 13

Figure 6. Depictions of the Do Not Enter Icon sign messages. 21

Figure 7. Depiction of the Do not Cross or Turn left Icon sign messages. 21

Figure 8. Depictions of the Proceed with Caution Icon sign messages. 21

Figure 9. A depiction of the hazard sign messages. 26

Figure 10. Order of screen presentation during timed comprehension task. 33

Figure 11. Angles that signs were placed in for Location set A at the near and far locations. 49

Figure 12. Confidence in identifying what traffic the signs were telling information about for the three angles over both sign types. 52

Figure 13. Agreement with how comfortable it was to view signs at the Pn for the three angles over both sign types. 54

Figure 14. Agreement with how comfortable it was to view signs at Pf for the three angles over both sign types. 54

Figure 15. Agreement with how easy it was to see the signs at the Pn for the three angles over both sign types. 55

Figure 16. Example of an image set (Icon sign viewed from Pn) shown to participants when they were asked to rank the angles in terms of how well they map information from the sign to the roadway conditions. 56

Figure 17. Frequency of top-rankings for the three angles over both sign types after viewing from Pn. 56

Figure 18. Frequency of top-rankings for the three angles over both sign types after viewing from Pn. 57

Figure 19. The angles for location sets A and B that will be used for the Location study. Range of angles for signs at location set A that were concluded to be easiest to read, lead to least amount of confusion, and were preferred by observers. 58

Figure 20. Location set options, where participant viewing locations are indicated by a white car labeled 'P'. 59

Figure 21. Accuracy of identifying traffic that the sign was giving information about for both location sets and genders. 64

Figure 22. Confidence in identifying what traffic the signs were telling information about for both location sets and genders. 65

Figure 23. Preference choices when at Pn (An v. Bn) and Pf (Af v. Bf). 69

Figure 24. Visual description of "gap", "lag" and "lead gap". 81

Figure 25. Means and standard deviations of accepted gaps and safety margins by sign condition for the near lanes when crossing from the stop sign. 83

Figure 26. Means and standard deviations of accepted gaps and safety margins by sign condition for the far lanes when crossing from the median. 83

Figure 27. Means and standard deviations of Wait Time for waiting at the stop sign and in the median. 85

Figure 28. Percentage of participants who made one-stage maneuvers per trial for each sign condition. 86

Figure 29. Usefulness and satisfying ratings for each sign. 88

LIST OF TABLES

Table 1. Design concepts for CICAS - SSA. 2
Table 2. Countdown sign original concepts and updated design options for testing. 4
Table 3. Icon and Hazard sign original concepts and updated designs for testing. 5
Table 4. Sample Demographics 7
Table 5. Rating scales for categorizing and scoring subject responses to the icons (from Campbell et al., 2004b). 11
Table 6. Overall comprehension ratings and appropriateness rankings for the countdown signs 16
Table 7. Comprehension ratings and appropriateness rankings for the countdown signs by age. 17
Table 8. Overall comprehension ratings for the Icon sign messages. 23
Table 9. Comprehension ratings for the Icon sign messages by age. 24
Table 10. Overall comprehension ratings for the hazard signs 27
Table 11. Comprehension ratings for the hazard signs by age 28
Table 12. Experiment two sample demographics 32
Table 13. Timed comprehension for the countdown sign's "Do Not Enter" messages. 35
Table 14. Timed comprehension results for the countdown sign's "Do not cross/turn left" messages. 36
Table 15. Timed comprehension results for the countdown sign's "Proceed with Caution" messages. 36
Table 16. Timed comprehension for the icon sign's "Do Not Enter" and "Do Not Cross or Turn Left" messages. 37
Table 17. Timed comprehension results for the icon sign's "Proceed with Caution" messages.. 39
Table 18. Timed comprehension results for the hazard sign's messages. 40
Table 19. Recommended design options for countdown stop-assist signs. 45
Table 20. Recommended icon designs. 46
Table 21. Possible correct responses to the question, "What traffic is the sign you just viewed telling you information about?" 51
Table 22. The number of positive and negative responses to the question, "Why did you prefer this pair of locations?" split by location set. 67
Table 23. The number of positive and negative responses to the question, "Why did you prefer this location?" for the near sign locations (An & Bn), split by location set. 69
Table 24. The number of positive and negative responses to the question, "Why did you prefer this location?" for the far sign locations (Af & Bf), split by location set. 70
Table 25. Summary of the differences between location sets A and B. 72

Table 26. CICAS-SSA interfaces. Each interface displays multiple messages depending on whether the driver is at the stop sign or in the median. 77

Table 27. Percentage of gaps rejected that were smaller than 7.5 s. 82

Table 28. Results for the 10-item post-drive questionnaire. 87

Table 29. Cohen's kappa for interrater reliability between the researchers. E-1

EXECUTIVE SUMMARY

In the United States it is recognized that crashes in rural areas are a cause for concern, especially crashes at rural intersections where inherent speeds may be associated with higher fatality rates (FHWA, 2004). Recent work has shown gap acceptance problems to be the key factor contributing to these crashes (Laberge, et al., 2006) as opposed to stop sign violation (Preston & Storm, 2003). However, the majority of intersection decision-support systems implemented at intersections have not attempted to provide specific information about the nature of available gaps in the approaching traffic or information that supports a driver's gap acceptance decision. In light of this, to reduce the crash risk at rural stop-controlled intersections, it has been recommended that intersection decision-support systems to assist drivers in responding to safe gaps be developed and deployed (Preston, Storm, Donath, & Shankwitz, 2004). The Cooperative Intersection Collision Avoidance System-Stop Sign Assist (CICAS-SSA) is an infrastructure-based driver support system that is used to improve gap acceptance at rural stop-controlled intersections. The SSA system will track vehicle locations on the major road and then display messages to the driver on the minor road.

The primary goal of the current work was to evaluate several candidate CICAS-SSA concepts in order to identify a single sign that may provide the greatest utility in terms of driver performance and usability at a real-world rural intersection. A secondary goal of the current work was to determine the ideal physical characteristics (i.e., location and rotation of a sign relative to drivers) of the candidate CICAS-SSA at a test intersection to maximize comprehension (and subsequent use) of the sign.

The primary goal was accomplished by conducting three studies. The first two studies examined icon use and word selection for three candidate CICAS-SSA signs. The conduct of these studies provided the justification and information needed for redesigning the three candidate signs. Results of this work indicated that prohibitive messages or messages that provided clear warnings resulted in the highest comprehension rates. In particular, an Icon sign produced the highest comprehension rates of all signs tested. The Icon's "do not cross/turn left" design had a much higher comprehension rate [40%] than the same state design for the Countdown sign [<10%].

The third study evaluated driving performance and usability for three candidate SSA sign designs compared to a baseline condition for the purpose of identifying the final candidate sign to be field tested at the Minnesota test intersection. The presence of CICAS-SSA signs affirms good decision making while also supporting drivers who may have difficulty in the selection and acceptance of a safe gap when crossing. For the Countdown and Icon signs, drivers reported using the CICAS-SSA information to help with their crossing decisions. However, performance data showed that the Countdown sign also resulted in behaviors that may elevate the risk of a crash, such as one-stage crossing maneuvers and misuse of the timer functionality. Although a similar distribution of unsafe gaps were rejected while using the Icon sign, participants chose gaps with larger safety margins and exhibited more two-stage crossing maneuvers with this sign compared to the Countdown condition. It is recommended that the Icon sign be implemented in an experimental field test.

The secondary goal was accomplished by conducting two studies that determined the optimal physical characteristics for the sign's location in order to maximize driver comprehension. Results of the work examining sign location indicated that a CICAS-SSA sign placed on the shoulder of the near-side road on the left side (for the driver positioned at the stop sign) along with a second sign located in the median in front and to the right of the driver (for a driver positioned in the median) was most preferred and resulted in adequate understanding. However, observations of sign locations at an actual intersection suggested that visibility of the signs may be poor and the potential of the signs to obscure expressway traffic was highly probable, especially for those drivers seated in larger vehicles (e.g., heavy trucks). In light of this finding, it was decided that for drivers at the stop sign, a CICAS-SSA sign is best positioned in the left-side median and that for drivers in the median, a CICAS-SSA sign is best positioned on the far right shoulder. Results of the study examining sign rotation angle indicated that a CICAS-SSA sign placed parallel to the mainline roadway was associated with a high degree of comprehension (i.e., drawing a clear association between the sign information and the roadway to which it applied); however, this angle also proved difficult to view. In contrast, a CICAS-SSA sign that was placed perpendicular to the minor roadway (i.e., directly facing a driver) was easy to view but was also associated with suboptimal comprehension. The 45 degree angle did not produce any errors in comprehension, was reported as comfortable and easy to view, and was preferred by over 75% of the respondents. Therefore, it was recommended that a 45 degree angle (or similar) be implemented in further testing.

1 Introduction

Rural intersection crashes more often result in fatalities because of the high speeds involved on rural highways (FHWA, 2004). In particular, intersections where a high-volume, high-speed multi-lane road is intersected by a lower-speed, lower-volume road controlled by a stop sign pose a problem due to the high speeds present on the main road and the need for drivers on the minor road to accelerate from a stop to enter this fast-moving traffic (see Figure 1). AASHTO recognized the significance of rural intersection crashes in its 1998 Strategic Highway Safety Plan (AASHTO, 1998) and identified the development and use of new technologies as a key initiative to address the problem of intersection crashes in Neuman, et al., 2003, Objective 17.1.4: "Assist drivers in judging gap sizes at Unsignalized Intersections." Previous research identified gap acceptance problems as a significant contributor to these crashes (see Laberge, Creaser, Rakauskas & Ward, 2006 for a review) as opposed to stop sign violation (Preston & Storm, 2003). To reduce the crash risk at rural stop-controlled intersections, recommendations have been made to develop and deploy intersection decision-support (IDS) systems to assist drivers in responding to safe gaps (Preston, Storm, Donath, & Shankwitz, 2004).

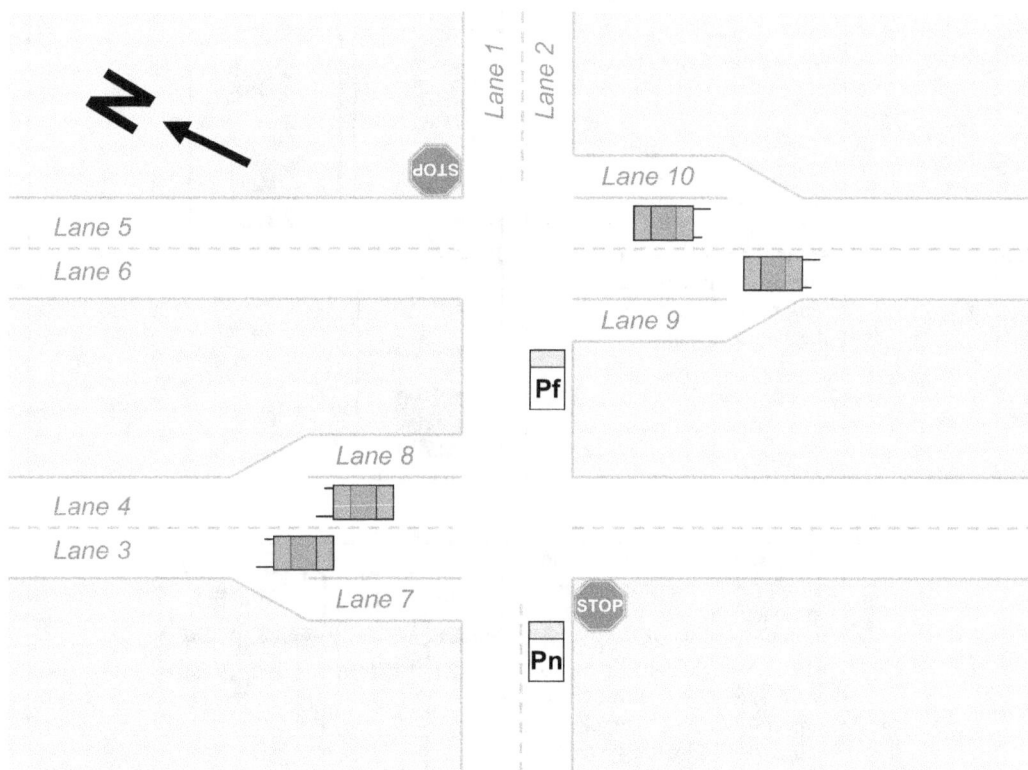

Figure 1. Diagram of a stop-controlled trunk-highway thru-stop intersection with relative location of SSA signs. Viewing locations while crossing the highway from the minor road are indicated by a white car labeled 'P'. The lane numbers correspond to the designations used by the Intelligent Vehicles Laboratory at the instrumented intersection of US 52 & Goodhue CR9. Consequently, lanes 3 & 4 constitute "Southbound" (SB) traffic and lanes 5 & 6 constitute "Northbound" (NB) traffic.

The initial Intersection Decision Support systems (IDS) studies that were conducted in the ITS Institute at the University of Minnesota identified the tasks required to cross a rural intersection and the most common driver errors at rural intersections (Laberge et al., 2006). This research then developed a list of information requirements for an infrastructure-based dynamic traffic sign to support drivers' gap acceptance at rural intersections. Several concept interfaces were initially tested using a driving simulator that directly replicated a test intersection from rural Minnesota and three of these concept interfaces were recommended for further study (see Table 1) in the Cooperative Intersection Collision Avoidance System – Stop Sign Assist (CICAS – SSA) project. The goal of the CICAS - SSA system is to track vehicle locations on the major road and then display messages to the driver on the minor road. Work completed by the Intelligent Vehicles Laboratory (Gorjestani et al., 2008) provided the information relevant to identify the warning algorithm for the SSA system.

Table 1. Design concepts for CICAS - SSA.

Sign Concept Name	Description	Example Images
Countdown Sign	This sign uses a timer countdown to show how far away approaching vehicles are in the nearest set of lanes. The icon on top provides a judgment about the safety of the available gap.	
Icon Sign	This sign uses icons and warning levels to depict approaching traffic. Prohibitive icons appear when a vehicle is too close to enter safely. A yellow icon appears when a vehicle is being tracked by the system.	
Hazard Sign	This sign flashes an alert to warn of approaching traffic when it is unsafe to enter the intersection.	

This report outlines a series of human factors studies conducted to identify the best SSA interface to be deployed in on-road field testing. As a goal, this series of studies sought to evaluate design issues related to each interface concept that could affect their comprehension and use by drivers in the real world. The main goals for these studies were to:

1. Determine the final interface design of each SSA concept interface by conducting standard comprehension testing used for traffic signs (HF 3.1 & 3.2 Comprehension Studies).
2. Determine the appropriate location and angle of rotation for the signs at the intersection to ensure high visibility and comprehension by drivers (HF 3.3 Location/Rotation Studies).
3. Identify the best concept for real-world deployment through evaluation of driver performance and usability of the candidate signs in a simulated testing environment that was designed to represent real-world driving at the test intersection (HF 3.4 Random Gap Simulator Study).

Overall, this series of experiments represents a comprehensive human factors review and testing of issues that affect drivers' abilities to use a CICAS – SSA interface safely. By evaluating the three candidate SSA interfaces systematically, the best candidate sign to be deployed for field testing can be identified and the data collected will allow insight into how drivers may respond and perform in relation to the SSA once it is deployed in the real world.

Comprehension Studies

It was important to evaluate the comprehension of the new SSA designs prior to final testing in the simulator because multiple design options arose during discussions on how to best design the signs to meet MUTCD standards and guidelines. New design options for each interface concept were derived from the results of the first simulator study (Creaser et al., 2007) and from recommendations made by the project's Technical Advisory Panel (TAP). A full description of the design issues is summarized in Appendix Q - "Summary of CICAS-SSA Functional Scope and Driver Infrastructure Interface (DII) Test Proposal" (which was complied by Janet Creaser and Nic Ward in 2007). Table 2 and Table 3 present the original concepts that were previously evaluated in the simulator (Creaser et al., 2007) and the newly designed options. Most changes involved the color and shape of icons, and the wording for text-based messages. The Countdown sign's redesign resulted in 2-3 design options for each message, while the Icon and Hazard sign redesigns resulted in only a single design to be tested for each message.

Two usability experiments were conducted to determine the final selection of icons and word usage for the Countdown messages and to verify the comprehensibility of the Icon and Hazard sign redesigns before moving onto a comparison between the three sign interfaces in the simulator study. The process of using standard usability testing meant that any differences in performance between the three SSA interfaces in the simulator study would more likely be due to the level of support provided by the interface and not poor design of the messages. Standard usability testing is also a cost-effective and efficient method for identifying the design of road signs.

Table 2. Countdown sign original concepts and updated design options for testing.

Sign Type	Message	Original Sign Concepts Tested in Simulator Study	Design Options Tested in Current Study			
Countdown Signs	Do Not Enter; Traffic Too Close	[DO NOT ENTER / VEHICLE WILL ARRIVE FROM THE LEFT IN / 3 SECONDS]	[VEHICLE FROM LEFT IN / 3 SECONDS]	[WAIT / VEHICLE FROM LEFT IN / 4 SECONDS]	[DO NOT ENTER / VEHICLE FROM LEFT IN / 5 SECONDS]	
	Do Not Cross/ Turn Left	[VEHICLE WILL ARRIVE FROM THE LEFT IN / 12 SECONDS]	[VEHICLE FROM LEFT IN / 11 SECONDS]	[VEHICLE FROM LEFT IN / 11 SECONDS]		
	Proceed with Caution	[CAUTION / VEHICLE WILL ARRIVE FROM THE LEFT IN / SECONDS]	[LOOK FOR TRAFFIC / VEHICLE FROM LEFT IN / SECONDS]	[LOOK FOR TRAFFIC / VEHICLE FROM LEFT IN / SECONDS]		

Table 3. Icon and Hazard sign original concepts and updated designs for testing.

Sign Type	Message	Original Sign Concepts Tested in Simulator Study	Design Options Tested in Current Study
Icon Signs	Do Not Enter; Traffic Too Close		
	Do Not Enter; Traffic Too Close		
	Do Not Cross/ Turn Left		
	Proceed with Caution; Car From Left		
	Proceed with Caution; No Cars Detected		
Hazard Signs	Do Not Enter; Traffic Too Close		
	Proceed with Caution; No Cars Detected		

It is important that all possible design options be evaluated by drivers of all ages. In particular, older drivers are at increased risk for crashes at rural intersections (Staplin & Lyles, 1991; Stamatiadis et al., 1991; Preusser et al., 1998) and are also more likely to misunderstand traffic signs (Shinar et al., 2003; Dewar, Kline, & Swanson, 1994). Therefore, it was important to test the comprehension of the different designs to ensure an appropriate final design would suit the needs of drivers of all ages.

The first experiment was a paper-and-pencil test that asked participants to explain, in their own words, what they thought each design option meant. The second experiment was a timed comprehension test where participants were allowed to view each design option for 1-3 s and were provided multiple choice responses for the meaning. Both methods are commonly employed for identifying the comprehension levels of icons and traffic signs (e.g., Campbell et al., 2004a; Chrysler et al., 2004). Because dynamic traffic signs are rare, the two methods used in these experiments have not been previously employed to examine comprehension of dynamic sign messages. Therefore, using both methods ensures a more comprehensive picture of how well drivers comprehend each design option which should result in the selection of the most appropriate icons and wording for each interface.

1.1 Experiment One Methods

1.1.1 Participants

Sixty participants were recruited for this study in three age groups, each consisted of 20 participants (10 male; 10 female). The age groups were Young (18-25), Middle (30-55) and Older(60+).

Table 4 shows the age and driving experience for each group. Although drivers were recruited by age to ensure diversity in the sample, the comprehension of drivers over age 60 was the main interest of this study. Therefore, results are reported by combining the Young and Middle age group results for comparison to the 60+ results. Participants were recruited through a local recruiting agency and paid $40 cash for their participation at the end of the study.

Table 4. Sample demographics.

Age Group	Mean Age (SD)	Mean Years Licensed (SD)	Mean Annual Mileage (SD)	Driving Frequency Past Month (N)
Young (18-25)	22.9 (1.8)	6.5 (2.3)	13405 (13656)	Never - 1
				Rarely - 1
				Sometimes - 4
				Most Days - 2
				Every Day - 12
Middle (30-55)	42.6 (6.7)	23.7 (8.7)	11968 (7090)	Never - 1
				Rarely - 0
				Sometimes - 3
				Most Days - 6
				Every Day - 10
Older (60+)	65.6 (4.5)	49.1 (1.2)	15960 (11654)	Never - 0
				Rarely - 0
				Sometimes - 0
				Most Days - 4
				Every Day - 16

1.1.2 Materials

The first experiment employed two paper-and-pencil tests that are based on work by Campbell et al. (2004a; 2004b). The first test is a comprehension test where participants view the sign and are asked to write down what they think it means. The second test is an appropriateness ranking test where participants are provided with the meaning of the sign and asked to choose between multiple design options by indicating which option most accurately represents the meaning presented. Campbell et al. recommends these two tests be used together to provide a better understanding of both comprehension and preference characteristics. An icon or sign with high comprehension and high preference is probably a good design.

Two test booklets were created for the paper-and-pencil comprehension and appropriateness ranking tests. The test booklets were identical, other than the presentation order of the signs. Signs were randomly ordered in Version 1 and then were presented in the reverse order for Version 2 to account for potential learning effects after viewing a number of signs. Appendix A presents Version 1 of the test booklet as an example.

Each booklet consisted of three parts. Part 1 contained a Driver Demographic Questionnaire. Part 2 contained all the test signs and participants were asked to write down what they thought each sign meant given the Driving Scenario described on the page. Because the Hazard sign is shown attached to the stop sign, it was decided to indicate which portion of the sign actually changed so that participants would not make incorrect assumptions about the other signs located with the stop sign (i.e., the "divided highway" and "one way" signs). Part 3 asked participants to rank the different design options for the Countdown signs according to how accurately the participant felt the sign portrayed a specific traffic scenario described on the page. For example, the description given for the Countdown test signs that were meant to indicate it was not safe to enter the intersection from the stop sign was "It is not safe to enter the intersection; traffic approaching from both directions. Traffic from left is 3 s away." Participants read this description and then were asked to rank the signs from 1 (most accurately represents intended message) to 3 (least accurately represents intended message).

1.1.3 Procedures

Participants completed the informed consent process upon arrival at the lab (see Appendix B for a copy of the informed consent document). The researcher provided an oral introduction to each participant at the beginning of the study (see Appendix C). This introduction outlined the tasks the participants would complete during the testing session. Participants were then provided with a Study Introduction sheet that described the context in which the signs would be used and showed a diagram of the type of intersection where these signs would be used. The context instructions for this study were written to describe a stop-controlled intersection that occurs on rural expressways when a smaller road (minor road) crosses a larger, multi-lane road (major road) with high-speed traffic. The written context provided for this study was developed using the guidelines in Campbell et al. (2004a) and is shown below. Too little or inappropriate context may result in unrealistically low comprehension while too much context may result in unrealistically high comprehension of the candidate signs. Pilot testing was used to finalize the descriptions and ensure that participants understood the signs and messages they would see were dynamic.

Figure 2 was included on the same page as the written context information to show participants the type of intersection to which the study referred. In addition to the study introduction, a video of a simulated intersection was shown to participants and a verbal description of the intersection given to further ensure they understood how traffic flowed at these intersections and what maneuvers were available to them as a driver sitting at the stop sign (i.e., turn left, turn right, cross over).

1.1.3.1.1 Study Introduction

"Our purpose is to investigate issues related to the use of active (or dynamic) traffic signs at rural intersections (see

Figure 2). Recent advances in technology have allowed the development of a system that can be placed near a STOP sign at a rural intersection to show drivers the state of traffic approaching the intersection on the main road. These signs are "smart" signs. This means the information on the sign changes in real time depending on the current traffic conditions near the intersection. This system presents information that helps you, the driver, make decisions about when to cross or turn at the intersection. The diagram below shows a typical rural intersection where a smaller road crosses a larger, multi-lane road with fast-moving traffic. The signs you will see today can be placed at or near the STOP sign to help a driver waiting at the STOP sign make a decision about when to enter the intersection."

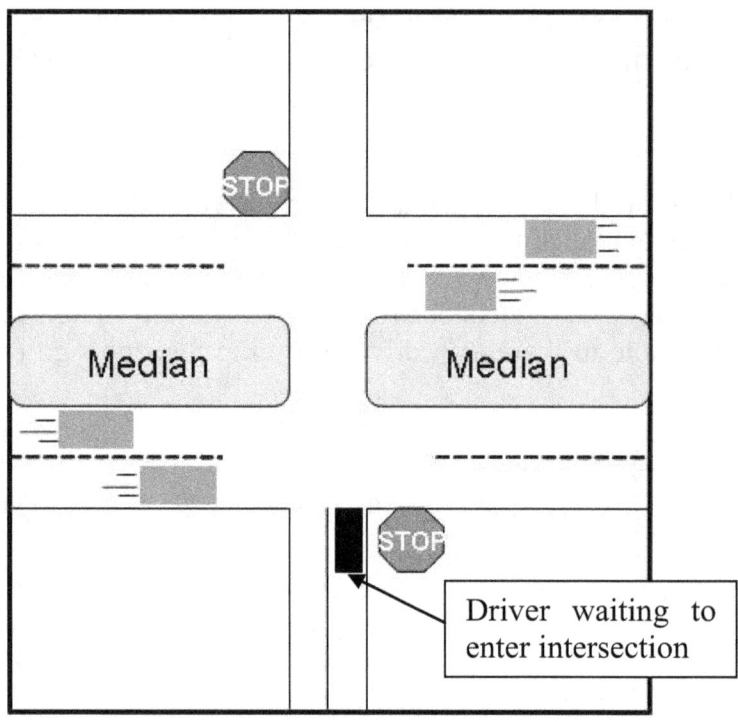

Figure 2. Diagram used in Study Introduction to show the rural, stop-controlled intersection.

Once participants finished reading the Study Introduction and viewing the video, the researcher answered any questions they had. After this, participants were provided with a test booklet. Half the participants received Version 1 while the other half received Version 2. Participants were then asked to complete the booklets on their own and to let the researcher know once they had completed all the questions. Participants were required to complete the booklet in order, starting with Part 1 (Demographic Questionnaire) and finishing with Part 3 (Ranking Appropriateness Questions). The researcher then reviewed the booklet to ensure all questions were answered. If a question was left blank, the researcher asked the participant to indicate why it was not completed. If it was simply missed, the participant was asked to complete the question. If the participant did not have an answer, they were instructed to write "do not know" as their answer. Once participants completed their booklets, the researcher thanked them for their time and provided remuneration for participation in the study.

1.1.4 Statistics

For the comprehension test participants' written responses were ranked on a 9-point scale of comprehension (see Table 5; Campbell et al., 2004b). A score of 1 or 2 indicates high comprehension of the sign's meaning. A score of 3 or 4 indicates a partial, low, understanding of what the sign means. Scores of 5-8 indicate responses that show no comprehension of the sign. Of significant importance is a score of 9, which indicates a response was a critical confusion or error. Critical confusions or errors occur when a participant's response indicates that they perceived the message to tell them to do something potentially unsafe (Campbell et al., 2004b). For example, if a lane departure warning was perceived by a participant to mean that they *should* leave the road rather than as a warning to *stay* on the road, it could have unintended, dangerous consequences for the driver. If an icon, symbol or sign has a high percentage of scores that are 1 and 2 but also a high percentage of critical confusions it is not a good design. A high percentage in of high comprehension scores (1 or 2) must be balanced with a low rate of critical confusions for a design to both well understood and safe.

In addition the comprehension results, the percentage of respondents who chose each design option for the appropriateness ranking is reported.

Table 5. Rating scales for categorizing and scoring subject responses to the icons (from Campbell et al., 2004b).

Comprehension Score	Description
1	The response matches the intended meaning of the icon exactly.
2	The response captures all major informational elements of the intended meaning of the icon, but is missing one or more minor information elements.
3	The response captures some of the intended meaning of the icon, but it is missing one or more major informational elements.
4	The response does not match the intended meaning of the icon, but it captures some major or minor informational elements.
5	The response does not match the intended meaning of the icon, but it is somewhat relevant.
6	Participant's response is in no way relevant to the intended meaning of the icon.
7	Participant indicated he/she did not understand the icon.
8	No answer.
9	For safety-critical icons, identify the number and percentage of critical confusions or errors. Critical confusions or errors reflect responses that indicate that the subject perceived the message to convey a potentially unsafe action.

1.2 Experiment One Results

1.2.1 Inter-rater Reliability

Two researchers rated the answers provided by all 60 subjects using the 9-point in order to ensure there were no inter-rater biases. Scoring criteria for each sign design was created prior to rating the signs, with determinants of major and minor informational elements included (see Appendix D for the scoring criteria for signs). To ensure that both researchers were familiar with how to apply the scale to the responses, they first scored the responses of 10 participants each and compared their results to calibrate themselves similarly to the rating scores. Both researchers then rated all 60 participants' responses for each sign.

The level of inter-rater reliability was determined using a consensus estimate approach. Consensus estimates of inter-rater reliability assume that reasonable observers can come to exact agreement on how to apply the various categories of a scoring system to the responses (Stemler, 2004). If two raters come to agreement on how to use a rating scale, they can be considered to share a common interpretation of the rating scale's construct. However, in this study, Critical Confusions were considered important to evaluate appropriately for the signs before assessing agreement. Therefore, all instances where a score of 9 was given to a response were examined first. If both researchers gave a response a score of 9, the result was kept. If one researcher gave a response of 9 while the other did not, these differences were resolved through consensus by reviewing each response and coming to agreement on whether the response reflected a critical confusion or error and which score should be applied. This ensured that the number of critical confusions or errors would be more accurately represented in the results. This was important because the CICAS-SSA system is meant to help the driver make safer crossing or turning decisions. Any chances for misinterpretation of a correct decision must be considered seriously.

The first approach to assess inter-rater reliability was to calculate the percent agreement among the two researchers. The rating scale has 9 scores where the adjacent scores reflect similar levels of comprehension. There are also clear boundaries between levels of comprehension, such that "high", "low", "no comprehension' and "critical confusions" are clearly delineated by the scale. Therefore, the researchers' scores for a response were considered to be the same if they were within +/- 1 score *and* were within the same comprehension level. Therefore, if one researcher applied a score of 1 and the other researcher applied a score of 2 to the same response, the scores were considered to agree because both indicate "high" comprehension. If one researcher applied a score of 3 and the other applied a score of 4 these were also considered to agree because both reflect "low" comprehension. However, if one researcher applied a score of 2 and the other applied a score of 3, these scores were considered to disagree because they are in different comprehension levels (i.e., high vs. low comprehension) even though they are within +/- 1 score of each other. Finally, scores that did not fall within +/- 1 of each other (e.g., 2 and 4 given) were considered to disagree. Calculating percent agreement using the +/- 1 method when rating scales have a wide range (e.g., 1-9 as opposed to 1-4) with adjacent scores being similar to one another is a popular modification to requiring full agreement for each level of the scale (i.e., only calculating percent agreement for ratings that match exactly) (Stemler, 2004). However, because the scale applied here had clear boundaries for levels of comprehension between certain rating scores, it was considered necessary to consider adjacent categories as disagreeing if they reflected different levels of comprehension.

Overall, there were 140 pairs of ratings out of 840 that did not meet the criteria for agreement. Therefore, the percent agreement for this study was 83.3% for researchers either giving the exact same score for a response or for giving scores that fell within +/- 1 in the same comprehension level. Overall, interrater reliability is considered good when consensus estimates of percent agreement are 70% or greater (Stemler, 2004). This result suggests that both researchers were reasonably reliable in applying the scores to the responses for all the signs. An additional analysis using Cohen's Kappa also showed that the experimenter ratings were significantly correlated (see Appendix E).

1.2.2 Comprehension Results

Because the scores applied to the responses were similar between the two researchers, the rates of comprehension presented in this section are taken from one researcher's scores only. The results are discussed by sign. All three sign types (Countdown, Icon, Hazard) were evaluated using the paper-and-pencil comprehension test. Additionally, the Countdown sign design options for each message were evaluated using the Appropriateness Ranking test.

Countdown Signs - There were three design options for the "do not enter" message (see Figure 3), two options tested for the "do not turn left/cross" message (see Figure 4) and two options tested for the "look for traffic" message (see Figure 5).

Figure 3. Depiction of the Do Not Enter message options.

Figure 4. Depiction of the Do Not Cross/Turn Left message options.

1.2.2.1.1.1

 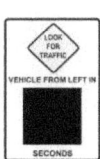

Figure 5. Depiction of the Proceed with Caution message options.

Table 6 presents the results of the comprehension ratings and the participant rankings for design preference for the Countdown signs. The "do not enter" (S2) message had the highest comprehension overall, with 58% of responses in the high comprehension level, 18% in the low comprehension level, 5% in the no comprehension level and 18% critical confusions. The design with the "wait" text and red hand (S3) had the next highest comprehension, with 43% high comprehension, 35% low comprehension, 3% no comprehension and 18% critical confusions. The design with the crash icon (S1) had the lowest comprehension, with 25% high comprehension , 42% low comprehension, 20% no comprehension and 13% critical confusions. Preference ratings were highest for the "wait" message (S3) at 45% and next highest for the "do not enter" message (S2) at 33%. Preference was lowest for the crash (S1) icon at 20%. Both S2 and S3 signs had similar rates of no comprehension and the same rate of critical confusions.

When comprehension was examined by age (see Table 7) for these message options the "do not enter" (S2) version had the highest comprehension rate overall for drivers aged 18-55 (58%) and for those 60+ (60%). The "wait" message (S3) showed the next highest comprehension for each group, at 38% for the 18-55 group and 55% for the 60+ group. Again, the crash icon (S1) showed the lowest comprehension, at 30% for the 18-55 age group and 15% for the 60+ group. Critical confusions were lowest for the crash icon (10%) and highest for the "wait" message (25%) for the 18-55 group. Critical confusions were lowest for the "wait" message (5%) and highest for the crash icon (20%) for the 60+ group. Overall, when looking at the comprehension levels and the critical confusion levels, the "do not enter" (S2) message is best for the 18-55 group while the "do not enter" or the "wait" messages (S3) are best for the 60+ group. Additionally, the 18-55 age group preferred the "wait" message (S3) similarly to the "do not enter" message (S2) (37.5% for both). However, the 60+ age group preferred the "wait" message (60%) more than the "do not enter" version (25%).

The comprehension rates were low for both options used to depict "do not cross/turn left". The percentage of responses for each sign that were in the high comprehension level was 7% for the diamond icon (S4) and 10% for the no-diamond message (S5). The low comprehension rates were also almost identical, with the diamond icon having 32% and the no-diamond message having 35% low comprehension. The percentage of critical confusions for both these messages was also high at 28% for the diamond icon and 35% for the no-diamond icon. Finally, the preference ratings for each message were also similar, with 45% of participant ratings the diamond icon as most appropriate and 55% rating the no-diamond message as most appropriate.

When comprehension rates for these message options are broken down by age group, they are similar across both age groups for both signs. Critical confusions are also fairly high for both age groups for each design option (>25%). Preference rankings were also similar for both age groups.

For the two options that were meant to convey the message "proceed with caution" (no cars detected in unsafe gap in either set of lanes), the sign with the diamond icon (S7) had 57% responses in the high comprehension level, 22% in the low comprehension level, 18% in the no comprehension level and 3% critical confusions. In comparison, the sign with the rectangle icon (S6) had a 27% high comprehension rate, a 48% low comprehension rate, a 17% no comprehension rate and an 8% critical confusion rate. However, the rectangular sign received significantly more first-place rankings than the diamond sign, with 77% of participants preferring it to the diamond sign.

When comprehension for these sign options is broken down by age group, the diamond version (S7) of the sign is best comprehended by both the 18-55 age group (63%) and the 60+ group (45%). Additionally, for both age groups critical confusions were lowest for the diamond version at 5% for the 18-55 age group and 0% for the 60+ age group. However, both age groups significantly preferred the rectangle version of the sign, with 75% of the 18-55 group and 80% of the 60+ group preferring it as the most appropriate depiction of the intended message.

Table 6. Overall comprehension ratings and appropriateness rankings for the countdown signs.

Message	Countdown Signs	Comprehension Ratings by Category									Overall Comprehension				Appropriateness Rankings		
		1	2	3	4	5	6	7	8	9	1-2 High	3-4 Low	5-8 None	9 Crit Con	1	2	3
Do Not Enter; Traffic Too Close	S1	13% (8)	12% (7)	17% (10)	25% (15)	8% (5)	8% (5)	3% (2)	0% (0)	13% (8)	25% (15)	42% (25)	20% (12)	13% (8)	20%	33%	47%
Do Not Enter; Traffic Too Close	S2	25% (15)	33% (20)	13% (8)	5% (3)	3% (2)	0% (0)	0% (0)	2% (1)	18% (11)	58% (35)	18% (11)	5% (3)	18% (11)	33%	35%	30%
Do Not Enter; Traffic Too Close	S3	15% (9)	28% (17)	13% (8)	22% (13)	3% (2)	0% (0)	0% (0)	0% (0)	18% (11)	43% (25)	35% (21)	3% (2)	18% (11)	45%	32%	23%
Do Not Cross/Turn Left	S4	2% (1)	5% (3)	20% (12)	12% (7)	7% (4)	13% (8)	13% (8)	0% (0)	28% (17)	7% (4)	32% (19)	33% (20)	28% (17)	45%	55%	
Do Not Cross/Turn Left	S5	2% (1)	8% (5)	13% (8)	22% (13)	8% (5)	8% (5)	3% (2)	0% (0)	35% (21)	10% (6)	35% (21)	20% (12)	35% (21)	55%	45%	

Table 7. Comprehension ratings and appropriateness rankings for the countdown signs by age.

Message	Sign	1	2	3	4	5	6	7	8	9	1-2 High	3-4 Low	5-8 None	9 Crit Con	1	2	3
Proceed with Caution	S6	17% (10)	10% (6)	17% (10)	32% (19)	7% (4)	7% (4)	0% (0)	3% (2)	8% (5)	27% (16)	48% (29)	17% (10)	8% (5)	77%	23%	
Proceed with Caution	S7	38% (23)	18% (11)	12% (7)	10% (6)	3% (2)	7% (4)	7% (4)	2% (1)	3% (2)	57% (34)	22% (13)	18% (11)	3% (2)	23%	77%	

			Comprehension Ratings by Category									Overall Comprehension				Appropriateness Rankings		
Message	Age	Countdown Signs	1	2	3	4	5	6	7	8	9	1-2 High	3-4 Low	5-8 None	9 Crit Con	1	2	3
Do Not Enter; Traffic Too Close	18-55	3	15%	15%	23%	25%	8%	5%	0%	0%	10%	30%	48%	13%	10%	25%	37.5%	37.5%
Do Not Enter; Traffic Too Close	60+	3	10%	5%	5%	25%	10%	15%	10%	0%	20%	15%	30%	35%	20%	10%	25%	65%
Do Not Enter; Traffic Too Close	18-55	5	25%	33%	10%	8%	3%	0%	0%	3%	20%	58%	18%	5%	20%	37.5%	25%	37.5%

	60+ Do Not Enter; Traffic Too Close	18-55 Do Not Enter; Traffic Too Close	60+ Do Not Enter; Traffic Too Close
	5 SECONDS	4 SECONDS	4 SECONDS
	25%	20%	5%
	35%	18%	50%
	20%	15%	10%
	0%	23%	20%
	5%	0%	10%
	0%	0%	0%
	0%	0%	0%
	0%	0%	0%
	15%	25%	5%
	60%	38%	55%
	20%	38%	30%
	5%	0%	10%
	15%	25%	5%
	25%	37.5%	60%
	55%	37.5%	20%
	15%	25%	20%

Table 7 continued from previous page.

Message	Age	Countdown Signs	Comprehension Ratings by Category									Overall Comprehension				Appropriateness Rankings	
			1	2	3	4	5	6	7	8	9	1-2 High	3-4 Low	5-8 None	9 Crit Con	1	2
Do Not Cross/ Turn Left	18-55		3%	3%	23%	15%	5%	13%	10%	0%	30%	5%	38%	28%	30%	40%	60%
Do Not Cross/ Turn Left	60+		0%	10%	15%	5%	10%	15%	20%	0%	25%	10%	20%	45%	25%	55%	45%
Do Not Cross/ Turn Left	18-55		0%	10%	18%	18%	10%	3%	3%	0%	40%	10%	35%	15%	40%	60%	40%
Do Not Cross/ Turn Left	60+		5%	5%	5%	30%	5%	20%	5%	0%	25%	10%	35%	30%	25%	45%	55%
Proceed with Caution	18-55		18%	13%	20%	28%	5%	5%	0%	3%	10%	30%	48%	13%	10%	75%	25%
Proceed with Caution	60+		15%	5%	10%	40%	10%	10%	0%	5%	5%	20%	50%	25%	5%	80%	20%

Table 7 continued from previous page.

Message	Age	Countdown Signs	Comprehension Ratings by Category									Overall Comprehension				Appropriateness Rankings	
			1	2	3	4	5	6	7	8	9	1-2 High	3-4 Low	5-8 None	9 Crit Con	1	2
Proceed with Caution	18-55		40%	23%	8%	10%	5%	8%	0%	3%	5%	63%	18%	15%	5%	25%	75%
Proceed with Caution	60+		35%	10%	20%	10%	0%	5%	20%	0%	0%	45%	30%	25%	0%	20%	80%

20

Icon Signs - Multiple design options of the icon sign were not tested during this study, thus there are no appropriateness rankings for this sign. The comprehension rates are for each potential message the sign could convey.

Overall, comprehension was high when the sign messages indicated that a driver could not enter the near or far lanes (S9; 62%) (see Figure 6 for a depiction of the sign) and when the configuration showed a driver could not enter the near lanes (S8; 65%) from the stop sign (see Table 8). These two messages were also highly comprehended by the 18-55 age group (58% & 70%, respectively) and the 60+ age group (65% for both messages; see Table 9).

Figure 6. Depictions of the Do Not Enter Icon sign messages.

Comprehension was slightly lower for the "do not cross/turn left" message (see Figure 7 for a depiction of the sign), with a high comprehension rate of 40%, a low comprehension rate of 40%, a no comprehension rate of 15% and a critical confusion rate of 5%. The comprehension of this message is higher than the same message set for the Countdown signs. Additionally, it had similar comprehension rates for both the 18-55 age (38%) group and the 60+ group (45%).

Figure 7. Depiction of the Do not Cross or Turn left Icon sign messages.

Figure 8. Depictions of the Proceed with Caution Icon sign messages.

The comprehension rates for the proceed with caution message was lowest, with a high comprehension rate of 20% for the "car from left" (S11) message and a 25% rate for the "no cars detected" (S12) message Figure 8. The 60+ group showed much lower comprehension (5%) for the "car from left" message compared with the 18-55 group (28%). This older group also showed lower rates of high comprehension (15%) for the "no cars detected" message compared to the 18-55 group (28%). Overall, critical confusions for these two messages were 7% for S11 and 2% for S12.

Table 8. Overall comprehension ratings for the Icon sign messages.

Message	Icon Signs	1	2	3	4	5	6	7	8	9	1-2 High	3-4 Low	5-8 None	9 Crit Con
Do Not Enter; Traffic Too Close	S8	30% (18)	32% (19)	20% (12)	5% (2)	2% (1)	5% (3)	0% (0)	0% (0)	7% (4)	62% (37)	25% (14)	7% (4)	7% (4)
Do Not Enter; Traffic Too Close	S9	22% (13)	43% (26)	18% (11)	3% (2)	3% (2)	3% (2)	0% (0)	0% (0)	7% (4)	65% (39)	22% (13)	7% (4)	7% (4)
Do Not Cross/Turn Left	S10	18% (11)	22% (13)	30% (18)	10% (6)	5% (3)	8% (5)	2% (1)	0% (0)	5% (3)	40% (24)	40% (24)	15% (9)	5% (3)
Proceed with Caution; Car From Left	S11	10% (6)	10% (6)	7% (4)	17% (10)	12% (7)	27% (16)	12% (7)	0% (0)	7% (4)	20% (12)	23% (14)	50% (30)	7% (4)
Proceed with Caution; No Cars Detected	S12	15% (9)	8% (5)	7% (4)	7% (4)	10% (6)	45% (27)	7% (4)	0% (0)	2% (1)	23% (14)	13% (8)	62% (37)	2% (1)

23

Table 9. Comprehension ratings for the Icon sign messages by age.

Message	Age	Icon Signs	Comprehension Ratings by Category									Overall Comprehension			
			1	2	3	4	5	6	7	8	9	1-2 High	3-4 Low	5-8 None	9 Crit Con
Do Not Enter; Traffic Too Close	18-55	S8	30%	28%	23%	5%	3%	5%	0%	0%	8%	58%	28%	8%	8%
Do Not Enter; Traffic Too Close	60+	S8	30%	40%	15%	5%	0%	5%	0%	0%	5%	70%	20%	5%	5%
Do Not Enter; Traffic Too Close	18-55	S9	25%	40%	20%	3%	3%	3%	0%	0%	8%	65%	23%	5%	8%
Do Not Enter; Traffic Too Close	60+	S9	15%	50%	15%	5%	5%	5%	0%	0%	5%	65%	20%	10%	5%
Do Not Cross/Turn Left	18-55	S10	20%	18%	38%	13%	3%	5%	3%	0%	3%	38%	50%	10%	3%
Do Not Cross/Turn Left	60+	S10	15%	30%	15%	5%	10%	15%	0%	0%	10%	45%	20%	25%	10%

Table 9 continued from previous page.

Message	Age	Icon Signs	Comprehension Ratings by Category									Overall Comprehension			
			1	2	3	4	5	6	7	8	9	1-2 High	3-4 Low	5-8 None	9 Crit Con
Proceed with Caution; Car From Left	18-55	DIVIDED HIGHWAY	15%	13%	10%	8%	15%	28%	5%	0%	8%	28%	18%	48%	8%
Proceed with Caution; Car From Left	60+	DIVIDED HIGHWAY	0%	5%	0%	35%	5%	25%	25%	0%	5%	5%	35%	55%	5%
Proceed with Caution; No Cars Detected	18-55	DIVIDED HIGHWAY	20%	8%	8%	10%	10%	38%	5%	0%	3%	28%	18%	53%	3%
Proceed with Caution; No Cars Detected	60+	DIVIDED HIGHWAY	5%	10%	5%	0%	10%	60%	10%	0%	0%	15%	5%	80%	0%

25

Multiple design options were also not tested for the Hazard sign. High comprehension for the "traffic too close" (S13) message was 68% while it was only 38% for the "no cars detected" (S14) message (see Table 10). Critical confusions were low for both messages at 5% for the "traffic too close" message (see Figure 9 for a depiction of this sign) and only 1% for the "no traffic detected" message. For the "traffic too close" message, high comprehension was 63% for the 18-55 group and 80% for the 60+ group (see Table 11). For the "no cars detected" message it was 38% for the 18-55 group and 40% for the 60+ group. Critical confusions were low for each age group, with the 18-55 age group having 8% for S13 and 3% for S14. The 60+ group had a 0% critical confusion rate for both messages.

Figure 9. A depiction of the hazard sign messages.

Table 10. Overall comprehension ratings for the hazard signs.

Message	Hazard Signs	1	2	3	4	5	6	7	8	9	1-2 High	3-4 Low	5-8 None	9 Crit Con
Do Not Enter; Traffic Too Close	S13	53% (32)	15% (9)	13% (8)	3% (2)	3% (2)	2% (1)	5% (3)	0% (0)	5% (3)	68% (41)	17% (10)	10% (6)	5% (3)
Proceed with Caution; No Cars Detected	S14	18% (11)	20% (12)	17% (10)	8% (5)	12% (7)	15% (9)	7% (4)	2% (1)	2% (1)	38% (23)	25% (15)	35% (21)	2% (1)

27

Table 11. Comprehension ratings for the hazard signs by age.

Message	Age	Hazard Signs	Comprehension Ratings by Category									Overall Comprehension			
			1	2	3	4	5	6	7	8	9	1-2 High	3-4 Low	5-8 None	9 Crit Con
Do Not Enter; Traffic Too Close	18-55		53%	10%	18%	5%	3%	0%	5%	0%	8%	63%	23%	8%	8%
Do Not Enter; Traffic Too Close	60+		55%	25%	5%	0%	5%	5%	5%	0%	0%	80%	5%	15%	0%
Proceed with Caution; No Cars Detected	18-55		18%	20%	13%	10%	13%	18%	5%	3%	3%	38%	23%	38%	3%
Proceed with Caution; No Cars Detected	60+		20%	20%	25%	5%	10%	10%	10%	0%	0%	40%	30%	30%	0%

1.3 Experiment One Conclusions

Overall, the results show that certain design options are favorable for the Countdown sign and that the Icon and Hazard sign designs are reasonably well comprehended, particularly for the prohibitive states. Because the sign designs follow a prohibitive framework, messages cannot explicitly tell drivers that it is "safe to go". Instead, the sign interfaces must indicate that traffic is not detected near the intersection and the driver must infer that a lack of prohibitive or warning information means they can probably enter the intersection if it is safe. These messages try to convey that some caution should be taken even though traffic is not detected so that drivers do not treat the decision-support messages they way the would a traffic light. The minimal information on these signs appeared to result in lower comprehension overall. For example, when the prohibitive icons are removed from the Icon sign designs to indicate "no traffic detected", comprehension for these signs was less than 25%. The only sign that performed well with an absence of information was the diamond icon with "Look for Traffic" in it for the Countdown sign (S7) with a 57% high comprehension rate. This sign indicated that no traffic was detected near the intersection, but that drivers should still take caution when entering. The Hazard sign suffered the same comprehension problem when the "traffic too close" text was removed (S14) with only a 38% comprehension rate compared to the 68% rate achieved with the active version of the sign (S13).

1.3.1 Countdown Messages

For the Do Not Enter message options, both the text "do not enter" and the "wait" with the red hand messages were well comprehended by the majority of drivers. Preference was slightly higher overall for the "wait" message both overall and for older drivers. Critical confusions were the same overall for these two messages. However, younger drivers had a much lower comprehension rate for this message when compared to the overall and the older driver comprehension rates.

Neither design option was well comprehended for the "do not cross or turn left" messages and there was a high rate of critical confusions for these messages. In the Countdown sign, the top icon (message) can apply to the whole intersection whereas the bottom timer only applies to the near lanes. It may be conceptually difficult for drivers to distinguish the difference in these two types of information, particularly in a paper-and-pencil test. For example, the comments below reflect confusion in trying to integrate the timer box with the "do not cross/turn left" (S4; S5) design options:

- *"It is okay to cross road as no approaching vehicles are indicated in black seconds box"*
- *"If any number is in the "seconds" box, it is unsafe to either go straight across or turn left at the intersection (by the car or the minor road) at that time. Presently, 0 seconds are displayed, the screen is blank so it is safe to proceed? Confusing."*

The low comprehension these "do not cross/turn left" (S4, S5) messages is likely due to the presence of two pieces of independent information on the sign. Drivers must infer that the only safe maneuvers are to turn right or proceed only to the median because traffic is not detected in the near lanes (or is outside unsafe threshold). It is likely that this sign cannot adequately be designed for easy comprehension without prior explanation of how the two message parts work together to provide an indication of the traffic flow on the major road. Appropriateness rankings for the "do not cross/turn left"(S4, S5) messages were also equally divided between the two options, further suggesting that neither design proved better at conveying the intended message. It is not clear how this sign could be redesigned to better convey the "do not cross/turn left" option.

For the "look for traffic (proceed with caution)" messages, the diamond icon had the highest comprehension overall and within each age group. However, participants overwhelmingly indicated that they felt the rectangular message was most appropriate was they knew what they message was intended to convey. This discrepancy is attributed to drivers' indicated preferences for the size of the rectangular message and the text inside when compared to the diamond. Although it was the case in this study that the rectangle was easier to read, the text requirements for the diamond sign would meet MUTCD requirements for legibility when placed at the intersection in its final form. It is a shortcoming of this experiment that legibility and sign color for both icons were not presented similarly, and was an artifact of how the images were generated. The main goal of this message is to instill a sense of caution in the driver so that they check the intersection before entering. Comprehension rates indicate that the diamond is the best design option.

There is a possible confound that could have influenced the comprehension rates of the two "look for traffic" designs during the comprehension portion of the experiment. The sign with the diamond icon does not have a time in the timer box while the one with the rectangle does. When the responses for both these signs are reviewed, the responses for the rectangle with the 10 s time in the timer box showed that participants may have more often misinterpreted the relationship of the icon to the time in the timer box. For example, they appeared to assume that the "look for traffic" message applied only to the traffic that was detected and that it did not imply, generally, that the driver should proceed with caution. In contrast, when the timer box was blank, participants appeared more able to understand that the "look for traffic" message meant to be cautious while crossing, even if traffic was not detected near the intersection. This confound is addressed in the second experiment by ensuring that both signs have a black timer box.

1.3.2 Icon Sign

For the Icon signs, the two signs that showed prohibitive information for the near lanes (S8) or the near and far lanes (S9) had the highest comprehension rates. The sign that prohibited entry into the far lanes (S10) had lower comprehension, but was still reasonably comprehended. The comprehension of these signs may be related to the direct mapping of familiar prohibitive information (i.e., the red circle and slash over the path indicators) onto a specific set of lanes for the intersection. This direct mapping of information onto a picture of the intersection adheres to principles of ecological interface design (Wickens, 1998). That is, the intersection display looks just like the layout of the roadway, with the changing pieces of information directly related to traffic on the roadway (i.e., yellow box representing a vehicle is farther from intersection while the red box is closer and presented in conjunction with other information indicating the danger of entering the intersection).

1.3.3 Hazard Sign

The Hazard Sign's comprehension rate for its active state ("traffic too close"; S12) is as high as the Countdown or Icon signs displaying this message. Again, this high level of comprehension is likely due to the text on the sign, which is a clear warning of what is happening at the intersection.

1.4 Experiment Two Methods

The second experiment examining sign comprehension employed a timed presentation of the sign design options. This method involves showing an image of the sign on the screen for a limited time and then asking participants to select the correct meaning or driving decision from a multiple-choice list. Chrysler et al. (2004) showed that a timed presentation method using a 3 s exposure produced similar comprehension rates of signs when compared to comprehension rates obtained through the use of an interactive driving simulator using the same signs. The authors suggested that the limited time presentation may produce a cognitive load similar to that observed in the simulator or on the road when drivers are required to identify and respond to signs while driving past them. They also discovered that unlimited time exposure may lead to overestimation of comprehension and thus recommended the time limited approach as a low-cost alternative to estimate comprehension. For this study, the timed presentation was modified to help estimate the minimum time required for most participants to correctly understand the meaning of the sign by presenting the signs repeatedly with different presentation times. The assumption is that a more comprehensible sign needs less viewing time to respond correctly. This method is different from the paper-and-pencil method in Experiment One, which allows participants as much time as they need to view a design option and answer each question.

1.4.1 Participants

Sixty participants were recruited for this study in three age groups with each group comprised of 20 participants (10 male; 10 female). The age groups were Young (18-29), Middle (30-59) and Older (60+). Table 12 shows the age and driving experience for each group. Although drivers were recruited by age to ensure diversity in the sample, the comprehension of drivers over age 60 was the main age interest of this study. Therefore, results are reported by combining the Young and Middle age group results for comparison to the 60+ results. Participants were recruited through a local recruiting agency and paid $40 cash for their participation at the end of the study.

Table 12. Experiment two sample demographics.

Age Group	Mean Age (SD)	Mean Years Licensed (SD)	Annual Mileage (N)	Driving Frequency Past Month (N)
Young (18-25)	24.3 (3.3)	7.2 (4.8)	<5000 - 3	Never - 1
			5001-10000 - 7	Rarely - 1
			10001-15000 - 7	Sometimes - 4
			15001-20000 - 1	Most Days - 2
			>20000 - 2	Every Day - 12
Middle (30-55)	42.0 (8.3)	26.8 (7.6)	<5000 - 2	Never - 1
			5001-10000 - 3	Rarely - 0
			10001-15000 - 7	Sometimes - 3
			15001-20000 - 2	Most Days - 6
			>20000 - 6	Every Day - 10
Older (60+)	63.9 (3.4)	44.7 (10.4)	<5000 - 2	Never - 0
			5001-10000 - 6	Rarely - 0
			10001-15000 - 6	Sometimes - 0
			15001-20000 - 3	Most Days - 4
			>20000 - 3	Every Day - 16

1.4.2 Apparatus

Participants were seated inside the HumanFIRST driving simulator and the sign options were presented to the participant using E-prime software (v1.1; Psychology Software Tools, 2003) running on an IMB-compatible PC running Windows XP. The simulator's front projector was used to display the images on the forward screen. Participant responses were logged using buttons on the Psychology Software Tools' Serial Response Box. The response box buttons were labeled to match the response options shown on the screen.

1.4.3 Procedures

Participants completed the informed consent process upon arrival at the lab (see Appendix B). The researcher provided an oral introduction to each participant at the beginning of the study that outlined the tasks participants would complete during the study. Participants were then provided with the same contextual information that was provided in Experiment One. Participants were seated in the simulator and the experiment was explained. All participants completed a practice session in which they were familiarized with the goals of the experimental tasks and were able to complete practice questions using the response box.

During the task, participants were first shown an instruction screen that explained how the task worked and what the goals of the task were. Once participants felt comfortable with the instructions they pressed the "start" button, which brought up the fixation screen. This fixation screen was a white screen with a cross in the middle that appeared for 3 seconds. This screen allowed participants to focus their attention on the screen and prepare themselves for the image presentation. The image screen appeared for either 1, 2 or 3 seconds depending on which time block the participant was completing. Time blocks were randomized for each subject and subjects saw all 14 images in each time block. Once the image had been presented for the appropriate amount of time, it was removed and the final screen appeared. The final screen showed participants the four possible response options that could apply to the sign message they had just seen.

A) Do not enter the intersection
B) Can enter the intersection to turn right only
C) Can enter the intersection to cross over, turn right or turn left
D) I do not know what the sign means

Participants were required to respond as accurately and as quickly as possible once the final screen appeared. The response box buttons were labeled to match the four possible answers to each question. Only one response was correct for each of the 14 sign images. Participants were instructed only to use the "don't know" option if they truly felt they could not answer the question based on their viewing of the sign image. The timed response was measured from when the response options screen appeared until participants entered a response on the response box. Once participants selected a response, it returned the participant to the instruction screen and they began the sequence again. Figure 10 shows the format for the Timed Comprehension Task. Once participants completed all the experimental tasks, they were thanked for their time and remunerated.

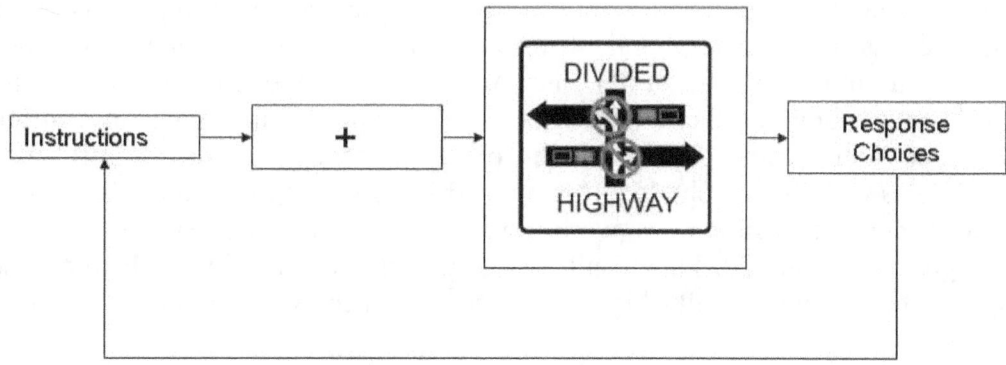

Figure 10. Order of screen presentation during timed comprehension task.

1.4.4 Statistics

For this experiment, two dependent variables are collected. First, the accuracy rates for each design option were presented as a percentage to indicate comprehension. Second, the average time to respond is calculated for each design option. Third, results indicate whether presentation time (1 s, 2 s, 3 s) had any effect on response times or accuracy rates.

1.4.5 Results

Overall, comprehension in this experiment was higher than in the first experiment. However, this is expected when the response options are provided for the participants as it provides context for the messages viewed. For all of the sign messages, the percentage correctly answered increased with each viewing, regardless of presentation time. This suggests that learning effects did occur during this experiment. It also indicates that drivers will become more familiar with the meaning of sign messages over repeated uses. For most signs, response time also became faster on subsequent viewings of a sign.

1.4.6 Countdown Sign

As with Experiment One, the "Do Not Enter" messages had the highest comprehension rate for this sign (see Table 13). Overall, the comprehension rates and response times were similar across the three messages. Comprehension rates were also significantly increased for the "Do Not Cross or Turn Left" messages (see Table 14). Both options had a comprehension rate of approximately 60% when participants were presented with a set of multiple choice options. This is a dramatic increase from the less than 10% comprehension rate for these two messages when participants were simply asked to describe what the message meant in the context of the smart sign description. However, both of these messages showed a large difference in comprehension between younger and older drivers, with older drivers showing significantly lower comprehension rates (approximately 45% for older drivers vs. approximately 80% for younger drivers). Finally, comprehension rates were approximately 66% for each of the "Look for Traffic" messages (see Table 15). This result is similar to that achieved with the diamond sign in Experiment One, but is substantially higher than the rate achieved for the rectangular sign in Experiment One.

Table 13. Timed comprehension for the countdown sign's "Do Not Enter" messages.

Message	Countdown Signs	Percent Correct [1]	Avg Resp Time	Age Effect [2]	Viewing Time Effect	Order Effect
Do Not Enter; Traffic Too Close	S1	85%	2.80 s	No	% correct: 1s < 2s & 3s RT: 1s & 2s faster 3s	• 1st viewing lowest % correct and slowest RT • 3rd viewing fastest RT • 2nd & 3rd viewing have similar % correct
Do Not Enter; Traffic Too Close	S2	83.9%	2.88 s	No	% correct: 1s, 2s, 3s similar RT: 2s & 3s faster than 1s	• 1st viewing lowest % correct and slowest RT • 3rd viewing fastest RT • 2nd & 3rd viewing have similar % correct
Do Not Enter; Traffic Too Close	S3	80%	2.98 s	No	% correct: 2s > 1s & 3s RT: 1s, 2s, 3s similar	• 1st viewing lowest % correct and slowest RT • 3rd viewing fastest RT • 2nd & 3rd viewing have similar % correct

1. Percent correct across all time viewings as scored by whether chosen response was correct.

2. Age effect is 'Yes' if the 60+ group had comprehension rates at least 15% lower than the Young or Middle groups.

Table 14. Timed comprehension results for the countdown sign's "Do not cross/turn left" messages.

Message	Countdown Signs	Timed Test				
		Percent Correct [1]	Avg Resp Time	Age Effect [2]	Viewing Time Effect	Order Effect
Do Not Cross/Turn Left	S4	59.4%	5.65 s	Yes Y=85%, M=53.3%, O=43.3%	% correct: 1s, 2s, 3s similar RT: 2s & 3s faster than 1s	• 1st viewing lowest % correct and slowest RT • 3rd viewing highest % correct and fastest RT
Do Not Cross/Turn Left	S5	60.6%	4.85 s	Yes Y=75%, M=56.6%, O=46.7%	% correct: 1s, 2s, 3s similar RT: 2s & 3s faster than 1s	• 1st viewing lowest % correct and slowest RT • 2nd & 3rd viewing have similar % correct and RT

1. Percent correct across all time viewings as scored by whether chosen response was correct.
2. Age effect is 'Yes' if the 60+ group had comprehension rates at least 15% lower than the Young or Middle groups.

Table 15. Timed comprehension results for the countdown sign's "Proceed with Caution" messages.

Message	Countdown Signs	Timed Test				
		Percent Correct [1]	Avg Resp Time	Age Effect [2]	Viewing Time Effect	Order Effect
Look for Traffic; Proceed with Caution	S6	65.6%	4.46 s	No	% correct: 1s, 2s, 3s similar RT: 3s faster than 1s & 2s	• 1st viewing lowest % correct and slowest RT • 3rd viewing highest % correct and fastest RT

Look for Traffic; Proceed with Caution	S7	66.1%	3.84 s	No	**% correct**: 1s < 2s < 3s **RT**: 1s, 2s, 3s similar	• 1st viewing lowest % correct and slowest RT • 2nd & 3rd viewing have similar % correct • 3rd viewing fastest RT

1. Percent correct across all time viewings as scored by whether chosen response was correct.
2. Age effect is 'Yes' if the 60+ group had comprehension rates at least 15% lower than the Young or Middle groups.

1.4.7 Icon Signs

Icon message comprehension rates were also significantly increased during this experiment compared with Experiment One. Comprehension rates for the "Do Not Enter" and "Do Not Cross or Turn Left" messages exceeded 80% for these messages, compared with rates ranging from 40% (S10) to 65% (S9) in Experiment One (Table 16). Of note for the Icon sign is that its "Do Not Cross or Turn Left" message retains a higher comprehension rate in this experiment compared with the Countdown sign's "Do Not Cross or Turn Left" options. Rates of comprehension for the "Proceed With Caution" messages were also higher in this study, but still lower than the "Do Not Enter" messages (see Table 17). Additionally, there was a difference in comprehension rates between younger and older drivers for sign S12.

Table 16. Timed comprehension for the icon sign's "Do Not Enter" and "Do Not Cross or Turn Left" messages.

Message	Icon Signs	Timed Test				
		Percent Correct[1]	Avg Resp Time	Age Effect[2]	Viewing Time Effect	Order Effect
Do Not Enter; Traffic Too Close	S8	81.1%	5.36 s	Yes Y=90%, M=78.3%, O=75%	**% correct**: 1s lowest; 2s & 3s similar **RT**: 3s fastest; 1s slowest	• 1st viewing lowest % correct and slowest RT • 3rd viewing highest % correct and fastest RT

Sign		% Correct	RT	Age Effect	Notes	
Do Not Enter; Traffic Too Close	S9	91.7%	2.93 s	No	**% correct**: 1s, 2s, 3s similar **RT**: 3s fastest; 2s slowest	• 1st viewing lowest % correct and slowest RT • 2nd & 3rd viewing have similar % correct • 3rd viewing has fastest RT
Do Not Cross/Turn Left	S10	83.9%	4.14 s	No	**% correct**: 1s, 2s, 3s similar **RT**: 2s & 3s faster than 1s	• 1st viewing lowest % correct and slowest RT • 3rd viewing highest % correct and fastest RT

1. Percent correct across all time viewings as scored by whether chosen response was correct.

2. Age effect is 'Yes' if the 60+ group had comprehension rates at least 15% lower than the Young or Middle groups.

Table 17. Timed comprehension results for the icon sign's "Proceed with Caution" messages.

Message	Icon Signs	Timed Test				
		Percent Correct[1]	Avg Resp Time	Age Effect[2]	Viewing Time Effect	Order Effect
Proceed with Caution; Car From Left	S11 DIVIDED HIGHWAY	64.4%	5.04 s	No	% correct: 1s, 2s, 3s similar RT: 1s, 2s, 3s similar	• 1st viewing lowest % correct and slowest RT • 2nd & 3rd viewing have similar % correct • 3rd viewing has fastest RT
Proceed with Caution; No Cars Detected	S12 DIVIDED HIGHWAY	75.6%	3.87 s	Yes Y=85%, M=75%, O=66.7%	% correct: 1s, 2s, 3s similar RT: 1s, 2s, 3s similar	• 1st viewing lowest % correct and slowest RT • 2nd & 3rd viewing have similar % correct • 3rd viewing has fastest RT

1. Percent correct across all time viewings as scored by whether chosen response was correct.

2. Age effect is 'Yes' if the 60+ group had comprehension rates at least 15% lower than the Young or Middle groups.

1.4.8 Hazard Sign

The rates of comprehension for the Hazard sign (Table 18 presents the comprehension results) did not change from Experiment One to this experiment. The rates were similar for both experiments. The "Traffic Too Close" message does not achieve the same comprehension rates as other "Do Not Enter" messages, suggesting that this message may not be perceived as a "Do Not Enter" message but rather simply an alerting message. The comprehension rate for the "Proceed With Caution" message for this sign is also considerably lower than the rates of comprehension for those messages in both the Countdown and the Icon signs.

Table 18. Timed comprehension results for the hazard sign's messages.

Message	Hazard Signs	Percent Correct [2]	Avg Resp Time	Timed Test		
				Age Effect	Viewing Time Effect	Order Effect
Do Not Enter; Traffic Too Close	S13	65.6%	4.07 s	No	% correct: 1s, 2s, 3s similar RT: 2s fastest; 1s slowest	• 1st viewing lowest % correct and slowest RT • 2nd & 3rd viewing have similar % correct and similar RT
Proceed with Caution; No Cars Detected	S14	36.1%	3.87 s	Yes Y=21.7%, M=53.3%, O=33.3%	% correct: 1s, 2s, 3s similar RT: 3s fastest; 2s slowest	• 1st viewing lowest % correct and slowest RT • 2nd & 3rd viewing have similar % correct and similar RT

1. Percent correct across all time viewings as scored by whether chosen response was correct.

2. Age effect is 'Yes' if the 60+ group had comprehension rates at least 15% lower than the Young or Middle groups.

1.5 Experiment Two Conclusions

Overall, comprehension for the sign messages was higher in this experiment than in the first experiment for both the Countdown and the Icon signs. The Hazard sign showed the same level of comprehension as Experiment One. As with Experiment One, comprehension for signs with explicit prohibitive information (e.g., "Do Not Enter") was better comprehended than those intended to mean "Proceed With Caution".

1.5.1 Countdown Sign

Unlike Experiment One, performance was best for the crash icon message when compared with the "Do Not Enter" and "Wait" messages for this experiment. However, the differences in performance were not large, with all three messages achieving an accuracy level between 80-85%. For the "Do Not Cross or Turn Left" messages both options performed better than in Experiment One. However, they also both performed similarly with approximately the same level of accuracy and response times overall and across age groups. Both of these messages still prove difficult for older drivers, where comprehension was quite a bit lower than for the younger age group. Finally, for the "Look for Traffic" messages, both options performed similarly in this experiment. There was no difference in accuracy rate and the diamond icon's average response time was only 0.6s faster than for the rectangle. In fact, the percentage correct for the rectangular "Look for Traffic" message was increased from 27% in Experiment One to 65.6% in Experiment Two whereas the diamond "Look for Traffic" message only changed from 57% in Experiment One to 66.1% in Experiment Two. The removal of the timer box confound may have improved performance for the rectangular "Look For Traffic" icon in Experiment Two.

1.5.2 Icon Sign

Comprehension rates were higher in Experiment Two than in Experiment One, particularly for the "Proceed with Caution" messages. There was a slight age effect for the "no cars detected" message, with older drivers having a lower comprehension rate than young and middle aged drivers. Response time was fastest for the "Do Not Enter" message when both lanes showed the prohibitive icons whereas response times were slower for the "Do Not Enter" messages when only one set of lanes showed the prohibitive icon. Participants may have had trouble matching a response when only one lane was indicated as prohibiting entry, particularly the far lane. When only the far lane shows a prohibitive icon, the participant must infer that it is safe to cross only to the median or turn right and select that response (e.g., "can enter the intersection to cross over, turn right or turn left"). In contrast, when viewing the message showing both lanes blocked, the information on the sign explicitly matches the information in the response option "do not enter the intersection".

1.5.3 Hazard Sign

The Hazard sign was the only sign not to show an improvement in comprehension rates when message response options were provided. The simplicity of this sign in supporting the gap acceptance task may mean that its messages are not well comprehended.

1.6 Discussion

Overall, comprehension of the design options was affected by the type of information used to present an intended message. Prohibitive or explicit warning information produced the highest comprehension rates, frequently in conjunction with fast response times. Messages that clearly conveyed a variation of "Do Not Enter" had the highest comprehension when compared to other possible messages an individual interface could present. Drivers are familiar with prohibitive traffic signs, which are more common than permissive signs, which may have aided comprehension rates and response times. In contrast, when a message required the driver to interpret what the absence of information meant (e.g., "No Traffic Detected") comprehension rates were low. This lower comprehension could be due to the lack of explicit information telling the driver what maneuvers are available to him or her. Instead, a driver must infer what the lack of prohibitive information means. Because liability issues require the sign interfaces to use a prohibitive framework, it is not possible to tell drivers when it is "safe to go". Instead, the messages are designed to indicate no traffic is detected. Therefore, the decision to go rests with the driver (e.g., "Look for Traffic").

1.6.1 Countdown Sign

For the Countdown sign, the high comprehension for the "Do Not Enter" and "Wait" messages in Experiment One may be due to the specific wording of each message. Participants frequently indicated the wording from the icon portion of these signs in their answers along with the timing information. However, participants rarely mentioned the crash icon information along with the timing information. Messages such as "Do Not enter" and "Wait" appear frequently in the course of everyday driving or road use in the form of "do not enter" signs and the "wait" hand for pedestrian crossings. Crash icons are not commonly presented on traffic signs. In Experiment One, participants indicated several meanings for the crash icon, such as that a crash happened near the intersection, that a truck specifically was going to hit them if they entered the intersection, or, more generally, that crashes were likely at the intersection (without reference to current conditions). These interpretations may increase participants' awareness of conditions at the intersection, but could also result in participants ignoring or misunderstanding the timer information and its relationship to current traffic conditions. When the multiple choice responses in Experiment Two were available to drivers, the crash icon had a much higher comprehension rate than in Experiment One. This indicates that providing drivers with the sign's intended message set may improve comprehension of the message set. Although educational campaigns have been suggested to improve comprehension of the SSA interfaces (Creaser et al., 2007), it is not possible to ensure everyone is familiar with the sign and its messages the first time they see it. Therefore, the best design options are those with high comprehension in both Experiments, such as the "Do Not Enter" and "Wait" message designs.

The main concern for the Countdown is the critical confusion rate, which was highest for these messages when compared to the Icon sign's "Do Not Enter" messages. Overall, critical confusion rates were highest for the "do not enter" (S1-S3) and "do not cross/turn left" (S4, S5) Countdown messages, regardless of comprehension rates. Rates of critical confusions were less than 8% for both the "look for traffic" design options. In comparison, critical confusion rates were below 7% for all Icon sign messages and both Hazard sign messages. There appeared to be two main reasons participants had a higher rate of critical confusions for the two Countdown sign messages ("do not enter" and "do not cross/turn left"). First, participants demonstrated some problems integrating the timer box and the icon to form a complete message. In the Countdown sign designs, the top icon (message) can apply to the whole intersection whereas the bottom timer only applies to only the near lanes. It may be conceptually difficult for drivers to distinguish the difference in these two types of information, particularly in a paper-and-pencil test. Second, several participants misinterpreted the timer box in this study by assuming that it would be safe to cross after the next vehicle passed (not necessarily true), or by assuming that traffic was approaching every "X" seconds. In addition to these two misinterpretations of the sign, many participants only phrased their answers for the Countdown design options in relation to the timer box and failed to mention the icon's meaning at all. This focus on the timer box is similar to what was noticed in the original IDS study, where drivers frequently reported using the timer information to make their crossing decisions but tended to ignore the icon on the top portion of the sign (Creaser et al., 2007).

1.6.2 Icon Sign

The Icon sign "Do Not Enter" messages also had the highest comprehension rates for this sign's message set. The message that prohibited entry into the far lanes ("Do Not Cross or Turn Left") had lower comprehension in Experiment One but showed improved comprehension when the response options were available in Experiment Two. The use of the red circle and slash is a common prohibitive icon used on traffic signs and comprehension may have been facilitated by its use. Additionally, the comprehension of these signs may be related to the direct mapping of familiar prohibitive information (i.e., the red circle and slash over the path indicators) onto a specific set of lanes for the intersection. This direct mapping of information onto a picture of the intersection adheres to principles of ecological interface design (Wickens, 1998). That is, the intersection display looks just like the layout of the roadway, with the changing pieces of information directly related to traffic on the roadway (i.e., yellow box representing a vehicle is farther from intersection while the red box is closer and presented in conjunction with other information indicating the danger of entering the intersection).

The lower comprehension rates for the Icon sign's "Proceed with Caution" messages may be due to the static presentation of the signs. For example, a driver viewing the Icon sign from the stop sign on a real roadway might be more likely to make the connection that the yellow vehicle icon on the Icon sign means a car is far enough away to cross, but that it is being tracked by the system and caution is required. This is because they would see both the yellow icon on the sign and the car in the distance approaching the intersection and might be better able to make the connection between the icon, the approaching vehicle and the level of caution required based on the car's visible distance from the intersection. Moreover, if they witness changes between the message states they may also be more likely to understand that it means no traffic is being tracked.

1.6.3 Hazard Sign

The Hazard sign's comprehension rate was reasonable for the "Traffic Too Close" message but the non-active state was not well comprehended. The presence of response options did not improve comprehension for either state suggesting that the Hazard sign may be more difficult than the other two signs to interpret in general. The Hazard sign supports fewer stages of the gap acceptance decision, and relies mostly on giving drivers an alert about oncoming traffic. Therefore, the decision to go rests more fully with the driver than it may in the more obviously prohibitive states of the Countdown and Icon signs.

1.6.4 Age

Age can be a factor the comprehension of traffic signs and symbols (e.g., Shinar et al., 2004; Dewar et al., 1994). In the previous IDS study older drivers (age 55+) had lower comprehension for all the sign concepts when compared to the younger age group (18-40) (Creaser et al., 2007). In this study, older drivers had problems with several of the messages. In particular, problems occurred with the Countdown signs "Do Not Turn Left or Cross" messages and with the "No Traffic Detected" or "Proceed with Caution" messages for the Icon and the Hazard signs. Older drivers may have a more difficult time inferring what they can do when there is no explicit prohibitive information on a sign than younger drivers do.

1.7 Conclusions

- Overall, prohibitive messages or messages that provided clear warnings resulted in the highest comprehension rates.

- Comprehension rates were lower when prohibitive information was absent on a sign and a cautionary approach to entering the intersection was the intended message of the sign.

- The presence of response options in Experiment Two improved comprehension rates for most messages.

- For the Countdown Signs, the "Do Not Enter", "Wait" and the "Look for Traffic" messages had the best overall comprehension.

- The Countdown sign had the highest rates of critical confusions. It may be that drivers find it difficult to integrate the fact that the top icon provides information about the entire intersection, while the bottom timer provides information about only one set of lanes.

- The Countdown's "Do Not Cross or Turn Left" message was poorly comprehended in Experiment One, but showed significant improvement in Experiment Two. The combination of the prohibitive message for the far lanes and the timer indicating no vehicles are detected in the near lanes may have resulted in confusion about the sign's meaning when response options were not available to participants.

- Overall, older participants had the most trouble with messages that were not prohibitive.

- Overall, the methods used in this study helped determine which designs were more easily comprehended by participants. However, providing appropriate context for a dynamic traffic sign using these methods is tricky. It is likely that comprehension (as evidenced by differences between Experiment One and Two) was affected because the signs were not displayed dynamically.

1.8 Recommendations

1.8.1 Countdown Sign

The Countdown sign had three message options for the "Do Not Enter" message. The results of both Experiments indicate that the "Do Not Enter" text in a yellow diamond is the best option (see Table 19). Although the "Wait" option performed similarly and was slightly more preferred by older drivers in Experiment One, it also showed lower comprehension among younger drivers in Experiment One. Younger drivers equally preferred both options.

There were two design options for the "Do Not Cross or Turn Left" message. Both options performed similarly in each experiment and neither was preferred significantly more than the other. However, the yellow circle and slash option performed slightly better than the diamond option and was slightly more preferred (see Table 19).

There were two design options for the "Look for Traffic" message. In Experiment One, the diamond icon option performed best, but was least preferred once drivers understood the intended meaning of the sign. In Experiment Two both options performed similarly well. The rectangular icon comprehension rate improved significantly in Experiment Two and was similar to the diamond icon. However, the diamond icon did not significantly improve its comprehension rate in Experiment Two. This suggests that the confound with the timer box noted in Experiment One may have affected the comprehension of the rectangular icon in Experiment One.

Overall, the three designs chosen for each message also have different icon shapes that may also cue drivers to message changes while interacting with the signs. The final design options in Table 19 reflect format changes based on engineering recommendations for the size of text and icons. The overall design concept and message sets remain the same.

Table 19. Recommended design options for countdown stop-assist signs.

Message	Recommended Design Options	Rationale
Do Not Enter	(DO NOT ENTER sign with "VEHICLE FROM LEFT IN 3 SECONDS")	• Highest comprehension rate in Exp 1 at 58%; 84% in Exp 2. • High preference (33%) • No age effect • Comprehension and response times improved with subsequent viewings in Exp 2

Message	Recommended Design Options	Rationale
Do Not Turn Left/Cross	(sign: VEHICLE FROM LEFT IN 8 SECONDS with curved arrows)	• Low comprehension in Exp 1 (10%); Moderate comprehension (61%) in Exp 2; higher than other alternative option in both experiments • Higher preference than other option (55% preference) • No age effect • Faster average response time than other option
Proceed with Caution	(sign: LOOK FOR TRAFFIC VEHICLE FROM LEFT IN ▄ SECONDS)	• Lower comprehension in Exp 1 (27%), but similar to other option in Exp 2 (66%). Difference in Experiment One may be due to a confound that was fixed in Experiment Two. • Highest preference (77%) • Comprehension and response times improved with subsequent viewings in Exp 2 • No age effect • Use of rectangle differentiates message from the other two sign messages that use diamond and circle

1.8.2 Icon Sign

Multiple design options were not tested for the Icon sign; however, the results of both experiments show good comprehension of all the messages. The final designs shown here reflect changes made based on engineering size requirements. The outlines for the yellow and red boxes were removed and the "divided highway" text was moved to the bottom of the sign to accommodate a larger design format for the lane outlines and icons. The overall design concept and message sets remain the same.

Table 20. Recommended icon designs.

Message	Recommended Design Options	Rationale
Do Not Enter	(sign with arrows and DIVIDED HIGHWAY)	• Retain original design with one modification: remove outlines of vehicle warning boxes • 65% High comprehension in Exp 1; 92% is Exp 2 • No age effect • Fast average response time (2.93 s) • Low critical confusions (7%)

Sign		Notes
Do Not Enter		• Retain original design with one modification: remove outlines of vehicle warning boxes • 62% High comprehension in Exp 1; 81% is Exp 2 – comprehension improves with subsequent viewings as well in Exp 2 • Slight age effect in Exp 2 • Average response time is initially slow (7.34 s on 1st viewing) but improves greatly with subsequent viewings (3.47 s on 3rd viewing) • Low critical confusions (7%) in Exp 1
Do Not Turn Left/Cross		• Retain original design with one modification: remove outlines of vehicle warning boxes • 40% High comprehension in Exp 1; 84% is Exp 2 • No age effect • Average response time (4.14 s) is faster than countdown signs overall and across viewings • Low critical confusions (5%) in Exp 1
Proceed with Caution		• Retain original design with one modification: remove outlines of vehicle warning boxes • 20% High comprehension in Exp 1; 64.4% is Exp 2 • Age effect in Exp 1 (Y+M=28%; O=5%) • Average response time is initially slow (8.47 s on 1st viewing) but improves greatly with subsequent viewings (3.24 s on 3rd viewing) • No critical confusions in Exp 1
Proceed with Caution		• Retain original design with one modification: remove outlines of vehicle warning boxes • 23% High comprehension in Exp 1; 76% in Exp 2 • Age effect in Exp 2 (Y=85%; M=75%, O=67%) • Average response time is similar to the Countdown options with same message • No critical confusions in Exp 1

1.8.3 Hazard Sign

The Hazard sign will be used in the simulator study in the same format that was tested here. The Hazard sign is only included in further testing as a baseline comparison for a simple warning versus multiple warning levels and judgments provided by the other two signs.

2 Rotation Study

The information on the SSA sign is displayed so that drivers can associate the information with traffic patterns around the intersection. There is a potential for confusion if the SSA signs are oriented or located in a way that the information presented can refer to more than one vehicle or traffic element. The findings from Creaser et al. (2007) suggested that the orientation and placement of the SSA signs may play a critical role in how the sign information is interpreted and employed to safely navigate an intersection. For this reason, the rotational orientation and location of the signs need to be examined so that the consistency of the mental model between the intersection traffic and sign are maintained.

To examine these issues, two studies were conducted in a simulator; one study examined the rotation of signs (the study report in this section) while a second study examined the rotation of the sign within a particular location (the study reported in the next section). In these studies, drivers observed the intersection from two locations on the minor road as if they were crossing the intersection (these locations are depicted as boxes labeled "Pn" for participant in the near position and "Pf" for participant in the far position in Figure 11).

As a result of previous evaluations (see earlier Comprehension section), two potential SSA signs remain viable concepts for implementation (Countdown and Icon signs). Both of these signs were of similar size and shape, differing in functionality, iconography, and information presented. Since the focus of the Rotation study is to identify the most effective rotational orientation of the signs, differences between the two sign types are not examined here and will be examined in the Random Gap Study (see Random Gap section below).

The rotation study was designed to determine if there was an optimal rotational orientation at which the signs could be placed that would be associated with improved comprehension. The results of this study are important because they will indicate the proper rotation of the final SSA sign candidates to be placed at the simulated and actual intersections for future evaluations (simulation and field test) and deployment relative to driver placement.

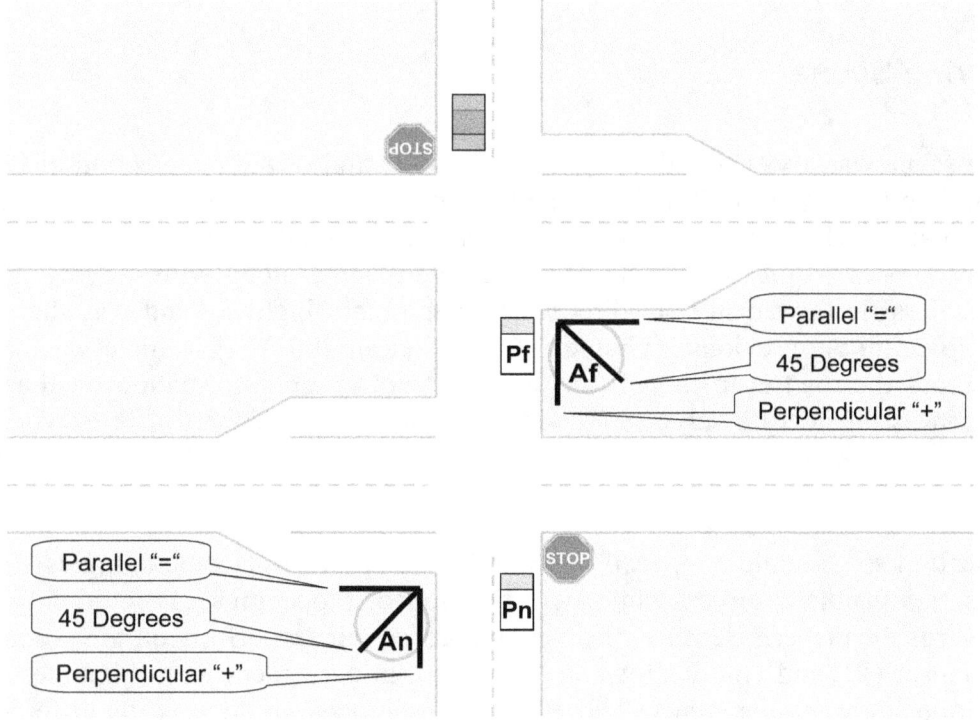

Figure 11. Angles that signs were placed in for Location set A at the near and far locations.

2.1 Methods

2.1.1 Participants

Seven females and three males (n = 10) (m = 27 years of age) participants were recruited from the University of Minnesota and surrounding metropolitan area.

2.1.2 Apparatus

A high-fidelity, limited-motion base driving simulator was employed to present virtual representations of the signs at the intersection. This immersive motion-base driving simulator is linked to a complete and full-sized vehicle and uses a five-channel 210-degree forward field of view with 1.96 arc-minutes per pixel resolution. Note that although this study employs the driving simulator and a naturalistic intersection crossing scenarios, they *do not* intend to present the drivers with control of the vehicle. Instead the simulator will be used as a visualization tool in order to present participants with views of the signs at various orientations at the intersection (naturalistic driving behavior will be tested in the Random Gap Study).

2.1.3 Driving Scene

The driving scene was a visual and topographical replication of the intersection of US Highway 52 and Goodhue County Road 9 in Goodhue County, Minnesota, where SSA signs are planned for implementation. The use of a virtual driving environment allowed for the display of signs at multiple locations in a quick, safe, and cost-effective manner along with moving traffic on the expressway. Traffic flowed in both directions on the main highway and was the same traffic stream employed in the previous IDS study (Creaser et al, 2007). A vehicle was parked at the stop sign across the road even though the signs did not report information on the minor road traffic. This was intended to serve as an additional possibility of the traffic to which the signs could refer.

Signs in location set A could be placed at a number of possible placement angles and still have their information visible to drivers while signs positioned at location set B could only be oriented directly towards the driver. Because of this, we focused our test efforts on the A locations only (Michael Manser (PI) and Ginny Crowson (AL) were in agreement with this assessment at the time of rotation study development). Therefore, the rotation study focused on different angles of signs in Location set A only.

2.1.4 Procedures

Upon arrival, participants read an instruction sheet detailing their activities relative to the experiment and then completed a human subjects' consent form (see Appendix B).

Each condition began with the experimenter instructing the participant to pay particular attention to the nearest sign. The experimenter also informed the participants that they would be observing this scene a number of times and will be asked questions about the driving scene and sign after each observation period. Only one state of the experimental signs was shown; for the Countdown sign the prohibitive icon and a low number in the timer was shown, while the Icon sign showed prohibitive red icons for both the near and far lanes. So that participants could compare the angles for each sign at each location, participants were shown the three angles at a particular location successively. For example, a participant viewing from Pn might see the Icon sign at the parallel, 45 degree, and then perpendicular angles during one condition.

Participants were then shown both the Icon and Countdown signs from the stop sign (Pn) and median (Pf) viewing locations, for a total of four conditions. The angle of the signs could be changed between three orientations: perpendicular, 45 degrees from perpendicular, or parallel to the expressway (see Figure 11 for diagram; see Figure 16 for depiction). The viewing locations and presentation were counterbalanced.

After observing the traffic scene and sign for 45 seconds, the experimenter asked the participant to indicate what traffic or vehicles the sign was referring to. Then the participant was asked to complete a page of ratings relating to the comprehension and usability of the signs at that particular angle (see Appendix F for Rotation Study questionnaires). After all angles at a location were viewed, participants were asked to rank the three viewing angles from best to worse and give reasons for their rankings.

2.1.5 Analyses

All measures were analyzed using a 3 (angle) by 2 (sign) repeated measures ANOVA with significant effects determined at $p < .05$, unless otherwise noted. Although sign type was included in our analysis model, our focus is on the main effects of rotation only. Therefore sign effects will only be discussed if they significantly interact with angle. For the comprehension and usability questions, separate analyses were conducted for responses while viewing from the stop sign (P_n) and median (P_f) viewing locations.

2.2 Results

Participant's performance was scored in terms of their comprehension of the signs and usability at each angle. Each measure was calculated by location and sign condition.

2.2.1 Comprehension

2.2.1.1 Accuracy in Mapping SSA Sign Information to Traffic Conditions

Table 21 presents correct responses for the question, "What traffic on US 52 is the sign giving you information about?" for both signs and viewing locations. All other responses were considered incorrect.

Table 21. Possible correct responses to the question, "What traffic is the sign you just viewed telling you information about?"

Location:	Sign:	
	Countdown	Icon
P_n	Southbound Southbound & Northbound	Southbound Southbound & Northbound
P_f	Northbound	Northbound Southbound & Northbound

For both sign types and locations, there were no errors when viewing signs at the parallel and 45 degree angles to US 52. Therefore, all the errors were made after viewing the signs at the perpendicular angle to US 52. There were 5 total errors (3 made by one participant), at least one per sign/location combination with two participants erring when viewing the Icon sign from Pn. Specifically, turning the sign to be perpendicular to US 52 made 3 participants think that the sign was referring to a vehicle at the stop sign heading West on CR9.

2.2.1.2 Confidence in Identifying Traffic that the Sign is Informing About

Participants were asked how confident they were in their identification of the traffic that the sign was telling information about (i.e. their confidence in the 3.2.1.1 accuracy measure). This was performed on a scale of "Not at all confident," "somewhat not confident," "Neutral," "Somewhat Confident," to "Completely Confident."

Pn. There were no significant differences between angle or sign types at the Pn location. Mean confidence overall was rated as 4.3. As shown in Figure 12 confidence was higher when viewing signs at the 45 degree angle ($M = 4.6$) when compared to the perpendicular angle ($M = 4.2$), $F(2,9) = 3.86$, $p = .040$ at the *Pf* location. Confidence was also marginally higher when viewing the Countdown sign ($M = 4.8$) when compared to viewing the Icon sign ($M = 4.0$), $F(1,9) = 4.53$, $p = .062$ at the *Pf* location.

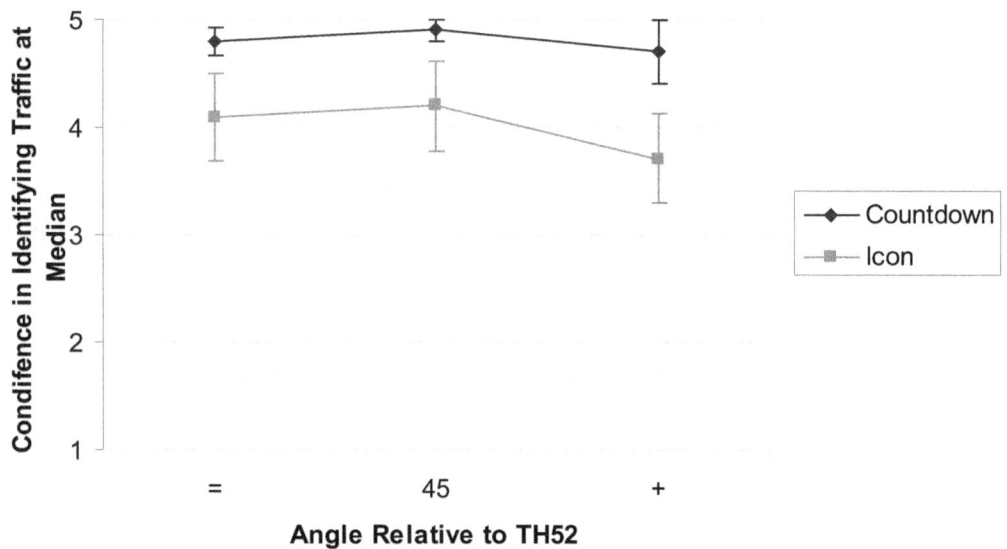

Figure 12. Confidence in identifying what traffic the signs were telling information about for the three angles over both sign types.

2.2.2 Usability

For all of the usability questions, participants were asked to rate their agreement with statements on a five point scale of "Strongly disagree," "Disagree," "Neutral," "Agree," to "Strongly Agree." See Appendix F for exact questionnaire wording.

2.2.2.1 Easy to Associate Information on the Sign to Traffic Conditions

Pn. There were no significant differences between angles. Mean agreement overall was rated as 3.1. Participants rated it easier to associate information while viewing the Countdown sign ($M = 3.6$) than while viewing the Icon sign ($M = 2.7$), $F(1,8) = 5.94, p = .041$.

Pf. There were no significant differences between angles. Mean agreement overall was rated as 3.3. Participants rated it easier to associate information while viewing the Countdown sign ($M = 3.9$) than while viewing the Icon sign ($M = 2.7$), $F(1,9) = 16.58, p = .003$.

2.2.2.2 Comfortable to View Sign in this Location

As shown in Figure 13, participants found it more comfortable to view the signs at the *Pn* location when viewing them at the 45 degree ($M = 4.3$) and perpendicular ($M = 3.9$) angles when compared to the parallel angle ($M = 2.6$), $F(2,18) = 26.63, p < .001$. As shown in Figure 14, participants found it more comfortable to view the signs at the *Pf* location when viewing them at the 45 degree angle ($M = 3.7$) when compared to the parallel angle ($M = 2.7$), $F(2,18) = 7.27, p = .005$.

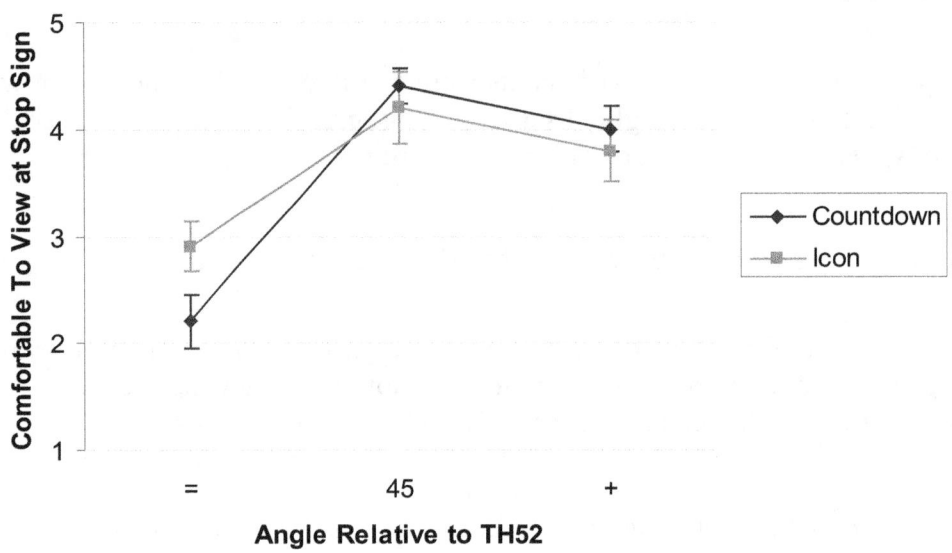

Figure 13. Agreement with how comfortable it was to view signs at the Pn for the three angles over both sign types.

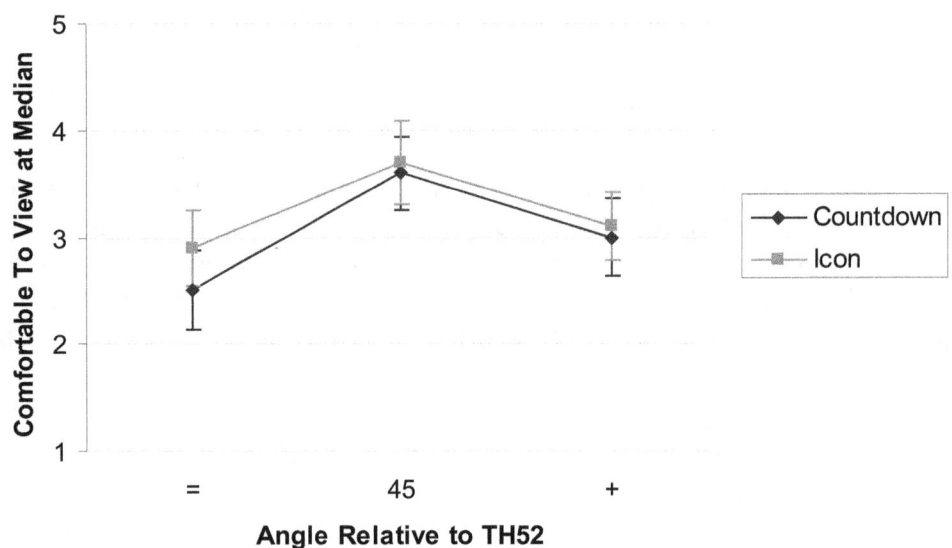

Figure 14. Agreement with how comfortable it was to view signs at Pf for the three angles over both sign types.

2.2.2.3 Obstructed View Approaching Traffic

There were no significant differences between angles at the Pn location. Mean agreement overall was rated as 1.9. There were no significant differences between angles at the Pf location. Mean agreement overall was rated as 2.1.

Easy To See at this Distance - As shown in Figure 15, participants found it easier to view the signs when viewing them at the 45 degree angle ($M = 4.4$) when compared to the parallel angle ($M = 3.8$), $F(2,18) = 5.19$, $p = .017$ at the *Pn* location. In addition, participants reported it was significantly less easy to see the Countdown sign ($M = 3.7$) than it was to see the Icon sign (M = 4.2), $F(2,18) = 3.92$, $p = .039$ at this location. There were no significant differences between angles at the Pf location. Mean agreement overall was rated as 3.3.

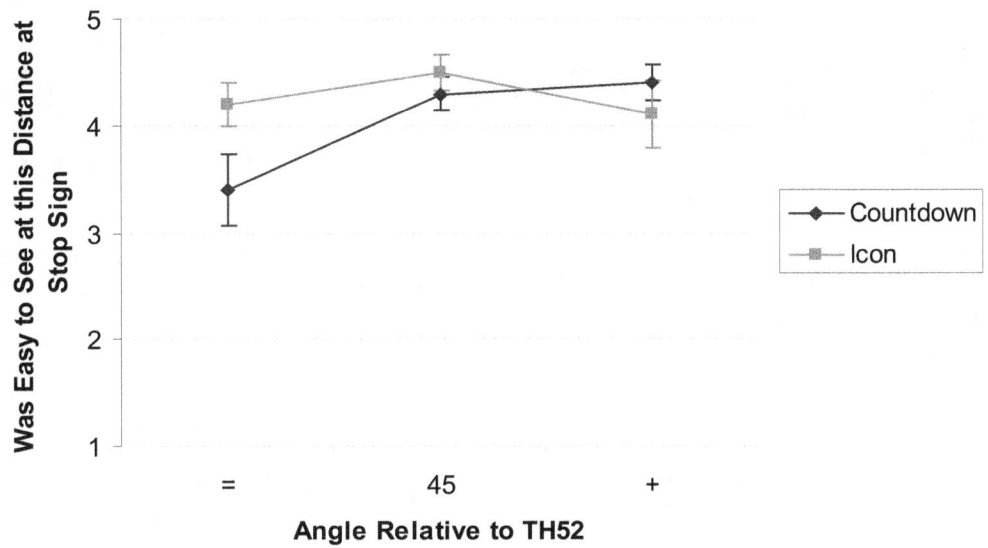

Figure 15. Agreement with how easy it was to see the signs at the Pn for the three angles over both sign types.

2.2.2.4 *Viewing Angle that Best Maps to Roadway*

Participants were shown images of the three rotation orientations they just observed (see Figure 16 for an example; see Appendix G for all examples) and were asked to rank the images in terms of how well they map information from the sign to the roadway conditions.

Figure 16. Example of an image set (Icon sign viewed from Pn) shown to participants when they were asked to rank the angles in terms of how well they map information from the sign to the roadway conditions. The images depict a sign parallel (=), 45 degree, and perpendicular (+) angle conditions.

As shown in Figure 17, participants ranked the 45 degree angle at location *Pn* most often (70% of all rankings) after viewing both sign types. The perpendicular angle was chosen 25% and parallel 5% of all rankings for this location. As shown in Figure 18, participants ranked the 45 degree angle at location *Pf* most often (80% of all rankings) after viewing both sign types. The perpendicular angle was chosen 20% and parallel 0% of all rankings at this location.

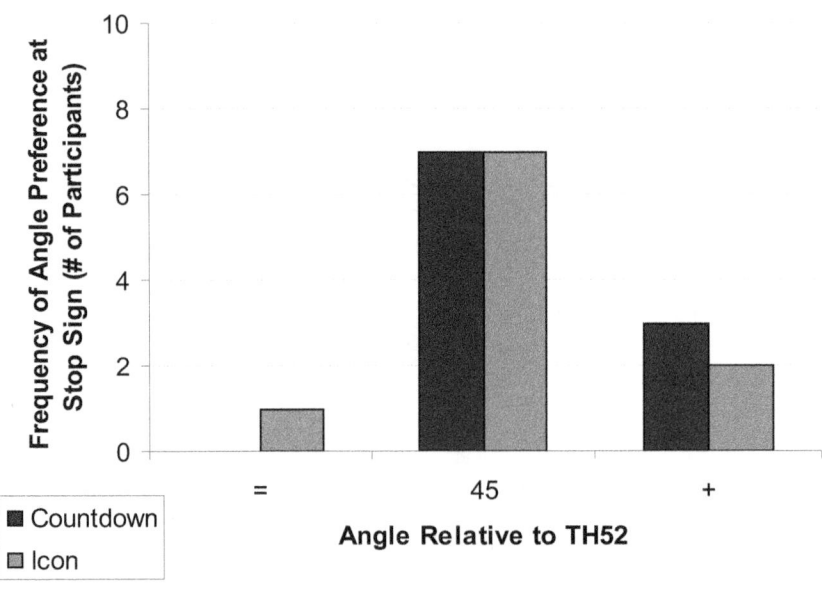

Figure 17. Frequency of top-rankings for the three angles over both sign types after viewing from Pn.

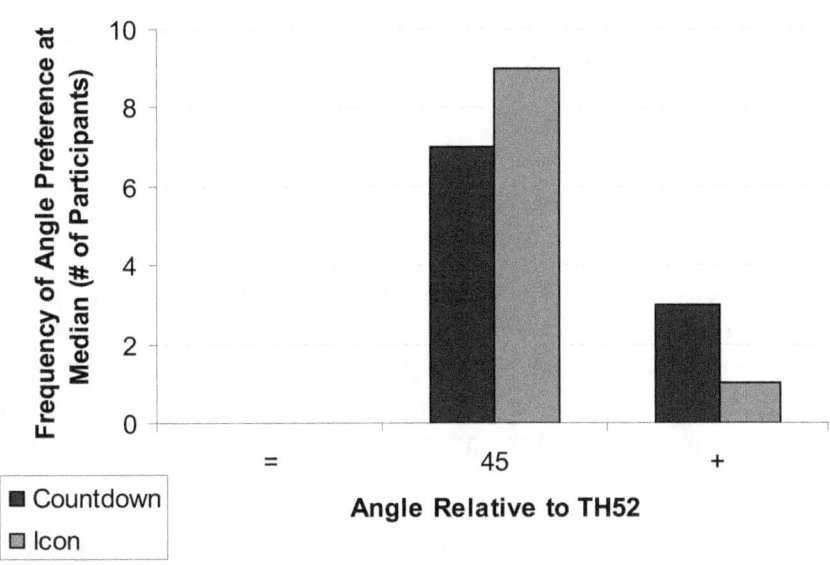

Figure 18. Frequency of top-rankings for the three angles over both sign types after viewing from Pn.

2.3 Discussion

It was important to examine the angular orientation of the SSA signs because the information on the signs is intended to be associated with traffic patterns around the intersection. By testing a number of angles at the respective viewing locations, the consistency of the mental model between the intersection traffic and sign was examined in this study.

Accuracy of the participant's mental map was quantified in terms of their ability to identify which traffic the signs were displaying information about. The comprehension results indicated that participants were able to accurately identify the traffic when the signs were placed at the parallel and 45 degree angles to US 52. However, participants made errors while viewing the signs perpendicular to US 52, which agrees with their confidence ratings in that they had the lowest confidence after viewing the signs placed at perpendicular angles at Pf.

When asked their opinions about the angles, there were no differences between angles for ease of associating information on the sign to traffic conditions, while at the Pn or Pf locations. At both locations, participants found it more comfortable to view the signs at a 45 degree angle than at the parallel angle.

At Pn, they also found it more comfortable to view the signs at a perpendicular angle than at the parallel angle. Participants reported that none of the angles obstructed their view of oncoming traffic at any location. While viewing from Pn, participants found it easier to view the signs at the 45 degree angle than at the parallel angle.

The countdown sign was reported as particularly less-easy to see at the parallel angle (potentially for not being able to read the text). At both locations, participants preferred the 45 degree angle the most while the perpendicular angle was also supported. There were a few differences between sign types, namely that Icon had lower confidence and association agreement than after viewing the Countdown sign.

2.4 Conclusions

The perpendicular angle was preferred by some participants and found it to be comfortable to view at the Pn location. However, the accuracy data suggests strongly that this angle should not be employed as it is the only angle that caused participants confusion. In contrast, the parallel angle did not produce any errors in comprehension. However, participants rated it as less comfortable and easy to view, and therefore suggests that this angle should also not be used.

The 45 degree angle did not produce any errors in comprehension, was reported as comfortable and easy to view, and was preferred by over 75% of the respondents. Therefore, it is recommended that a 45 degree angle (or similar) be implemented. However, the comprehension data suggested that signs closer to parallel may cause confusion. For this reason, it is also recommended that if a sign is to be angled, it should not exceed 45 degrees from parallel to the road it is giving information about (in this case, highway US 52), as depicted by the green curve in Figure 19.

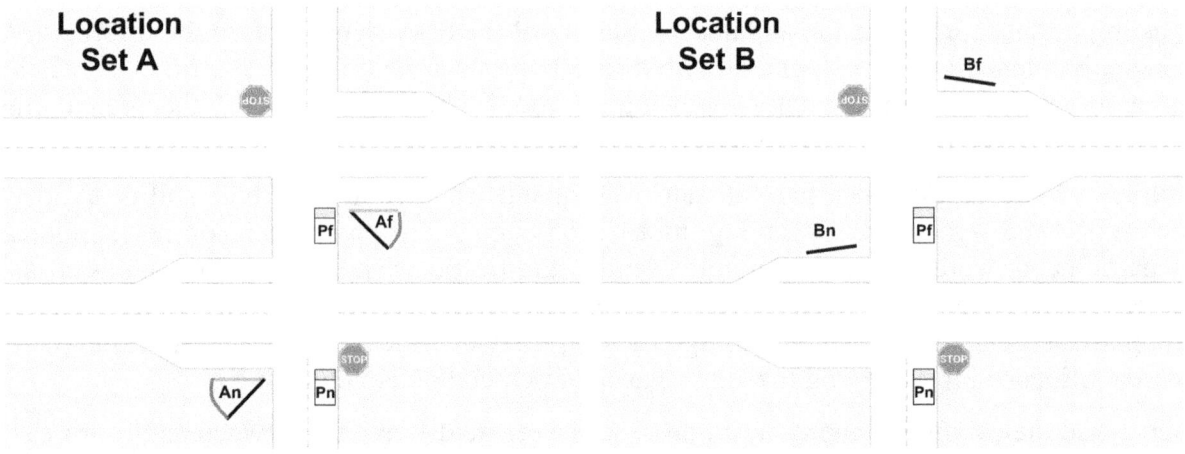

Figure 19. The angles for location sets A and B that will be used for the Location study. Range of angles for signs at location set A that were concluded to be easiest to read, lead to least amount of confusion, and were preferred by observers. The angles for signs at location set B are the only angles reasonable to be viewed from the Pn and Pf viewing locations.

3 Location Study

The information on the SSA sign is displayed so that drivers can associate the information with traffic patterns around the intersection. There is a potential for confusion if the SSA signs are oriented or located in a way that the information presented can refer to more than one vehicle or traffic element. As was just reported from the Rotation study, drivers are better able to associate information from the sign to traffic conditions when the signs are placed at rotational orientations that do not exceed 45 degrees from parallel to the road they are giving information about. It is now time to consider how the location of signs may also affect comprehension so that the consistency of the mental model between the intersection traffic and sign are maintained.

Previous evaluations (see earlier Comprehension section) have left two potential SSA signs as viable concepts for implementation (Countdown and Icon signs). Both of these signs were of similar size and shape, differing in functionality, iconography, and information presented. Since the focus of the Location study is to identify the most effective location of the signs, differences between the two sign types are not examined here and will be examined in the Random Gap Study (see Random Gap section below).

In the Location study, drivers observed the intersection from two locations on the minor road as if they were crossing the intersection (these locations are depicted as boxes labeled "P" in Figure 20). The experimental signs were grouped into two location sets (location set A and B) as shown in Figure 20.

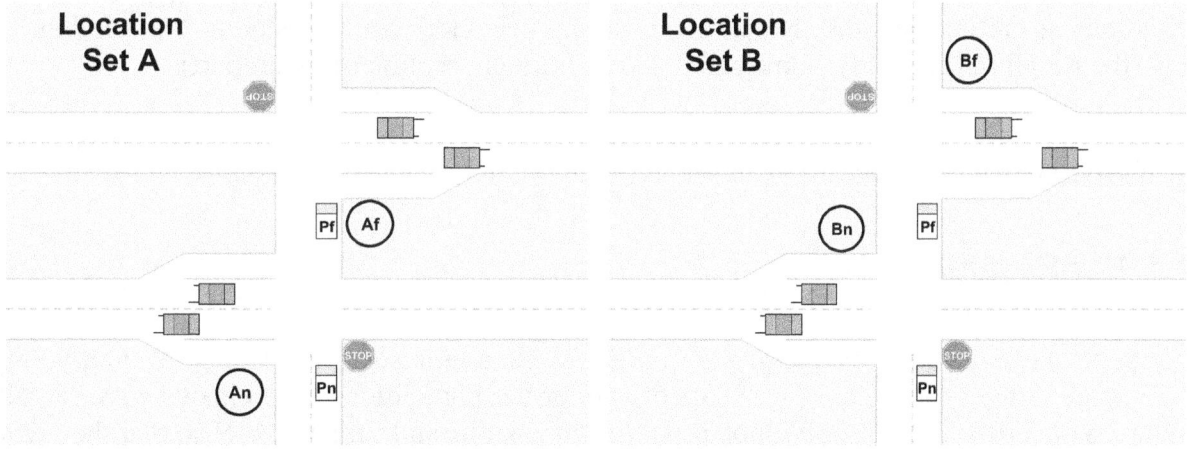

Figure 20. Location set options, where participant viewing locations are indicated by a white car labeled 'P'. Circles indicate sign locations labeled by set ('A' or 'B') and relative placement to a driver entering the intersection from the minor road ('n'ear or 'f'ar).

Location set A consists of a sign directly to the left of the driver (An) while at P_n and another sign directly to the right of the driver (Af) while at P_f. Location set B consists of a sign across the close lanes of traffic placed in the median (Bn) while at P_n and another sign across the far lanes of traffic placed on the opposite side of the road (Bf) while at P_f. Sign location sets were never mixed so that cross-combinations such as An & Bf or Bn & Af were not tested. Although the project team had previously agreed that location sets A and B were to remain as sets, subjective questions were added to the Location study methodology relating to individual locations in order to determine if a more logical combination set could be produced.

Although location set A was tested in the original IDS study (Creaser et al., 2007), there is no empirical data to support the utility of location set A over other possible locations (e.g., location Set B). Therefore it was necessary to test the utility of alternative locations with users in a simulator environment.

This study was designed to determine if there is a location set (location set A vs. location set B) which will provide higher comprehension of the SSA sign information. Comprehension will be quantified in terms of the participant's ability to match information on the signs to traffic conditions. The results of this study will be used to determine where the Countdown and Icon signs will be placed at the intersection during the following simulation studies and for the validation study field test.

Note that although these studies use the driving simulator and a naturalistic crossing situation, they *do not* intend to present the drivers with control of the vehicle. Instead the simulator will be used as a visualization tool in order to present participants with views of the signs at various orientations at the intersection. Naturalistic driving behavior will be tested in the Random Gap Study (the Random Gap study is presented in a subsequent section of this report).

3.1 Methods

3.1.1 Participants

Participants were recruited by age group in order to get a representative sample of both younger (18-35 years of age) and older ($\geq$ 60 years of age) driver populations. Participants were recruited from areas outside the I-494/I-694 loop around Minneapolis and St. Paul, MN so that they would be more familiar and experienced with rural thru-stop intersections. Fifty participants were recruited, of which six were excluded for not completing the experimental protocol as instructed, as determined by a post-experimental interview. Forty-four valid data sets balanced by gender were analyzed (young $n = 25$, $M = 28$ years of age; older $n = 19$, $M = 66$ years of age).

3.1.2 Procedures

Each participant completed an informed consent form prior to beginning the study (Appendix B). Participants again used the driving simulator (see description in the Apparatus section of the Rotation study, above) but did not drive, instead they were asked only to observe the driving scene which included vehicular traffic on the expressway. They were given three practice sessions in order to orient themselves with the simulated environment and with the experimental procedure. Participants experienced 16 counterbalanced experimental sessions (2 sign location sets x 2 sign types x 2 viewing locations x 2 exposures). During each location set, observers were expected to view the signs sequentially while waiting at the stop sign ("n"ear viewing location, P_n) and then while waiting in the median ("f"ar viewing location, P_f) as would be the case in the real world.

During each session, the driving scene would initially appear without SSA signs. Participants signaled the start of their observation task by pressing the brake pedal, at which time the dynamic SSA signs appeared in the virtual world and functioned as they would in the real world relative to traffic on the expressway. Participants were asked to observe the driving scene and report, "Which traffic is the sign giving you information about?" When participants had a response, they pressed the accelerator pedal and spoke their response out loud. In order to focus on initial reactions to each sign location/viewing location condition, only accuracy and timed responses from a participant's first exposure were examined.

After each driving scene, participants responded to a number of scaled and subjective usability questions (see Appendix H). After their first exposure to each sign location/viewing location/sign condition they were asked how confident they were in their response; after their second exposure they were asked the other usability questions. All questions were followed by space to explain their answers to the usability questions; summaries of these open-ended responses are included where applicable. After all conditions were completed, participants completed another questionnaire asking them their preference for sign location set and of particular signs locations separate from the location sets (see Appendix I).

3.1.3 Analyses

All measures were analyzed using a 2 (location set) by 2 (age group) by 2 (gender) repeated measures ANOVA with significant effects determined at $p = .05$, unless otherwise noted. Although age and gender were included in our analysis model, our focus is on the main effects of location only. Therefore age and gender effects will only be discussed if they significantly interact with location set. For the comprehension and usability questions, separate analyses were conducted for responses while viewing from the stop sign (P_n) and median (P_f) viewing locations (see Appendix H).

3.2 Results

Although sign, age, and gender were included in our analysis model, it is our intention to focus on the main effects of location. As such, these factors will only be discussed if they significantly interact with location.

3.2.1 Comprehension

Results from Pn at the stop sign (near) and "median" (far) viewing locations were analyzed separately.

3.2.1.1 Timed Comprehension Response Behavior (Response Time; RT)

Participants indicated with the brake when they were ready to begin the trial and with the accelerator when they thought they could identify the traffic that the sign was giving them information about. This time difference constituted their response time on the task.

Participants responded in this manner for all trials. An analysis comparing the location sets for their first response only on their first trial showed no significant differences between location set A and B at both viewing locations (both $p > .169$). This suggests that there are no differences between the location sets in terms of the time it takes participants to understand the signs when the situation is completely novel (i.e. a walk-up-and-use scenario).

To further explore the possibility of differences between location sets for response time, data from the first trial of each location set/sign exposure was compared by viewing location. There was no significant difference for location set at both viewing locations (both $p > .334$).

3.2.1.2 Accuracy - Ease of Mapping Information to Traffic Conditions

After indicating their timed response behavior, participants were asked to respond out loud what traffic they thought the sign was giving them information about. Table 21 (in the Rotation study results) presents the accepted correct responses for both signs and viewing locations. All other responses were considered incorrect.

(Participants were more likely to have more correct answers while viewing from the median and when viewing the countdown sign due to two potential conflicts in the implementation of this measure: A.) The countdown signs explicitly state that they refer to traffic coming from the right or left. This may have aided many drivers to give the right response while viewing these signs.

While at the Pn position it also may have made them ignorant that the sign was giving them information about the Northbound (far) traffic stream. B.) The traffic stream itself was such that Southbound traffic was consistently heavy for most of the trials. Because of this, participants many not have seen the information for Southbound traffic change during the trial, and they might have falsely thought the signs only gave them information about the Northbound (far) traffic stream.)

Participants responded in this manner for all trials. An analysis comparing the location sets for their responses only on their first trials of each location set/sign exposure showed no significant differences between location set A and B at both viewing locations (both $p > .218$). This suggests that there are no differences between the location sets in terms of the accuracy in understanding the signs when the situation is completely novel (i.e. a walk-up-and-use scenario).

To further explore the possibility of differences between location sets for response accuracy, data from all (both first and second) exposures of each location set/sign were compared by viewing location. Since no differences were found between the first and second trial at either viewing location (both $p < .331$), results for both exposures were combined for the following discussion.

Pn. While waiting at the stop sign, participants made less errors when the signs were placed in location set A ($M = 6\%$) when compared to location set B ($M = 9\%$) over both sign types, $F(1,40) = 4.11, p = .049$. There was also a significant main effect for sign type, $F(1,40) = 12.14$, $p = .001$, and a significant interaction between location and sign type, $F(1,40) = 4.11, p = .049$. As shown in Figure 21, participants made less errors while viewing the icon sign from the A locations in comparison to the B locations, which all observers correctly identified the traffic while viewing the countdown signs.

Pf. While waiting at the median, there was no difference in participants' error rate between the two location sets ($p = .331$; overall $M = 1\%$).

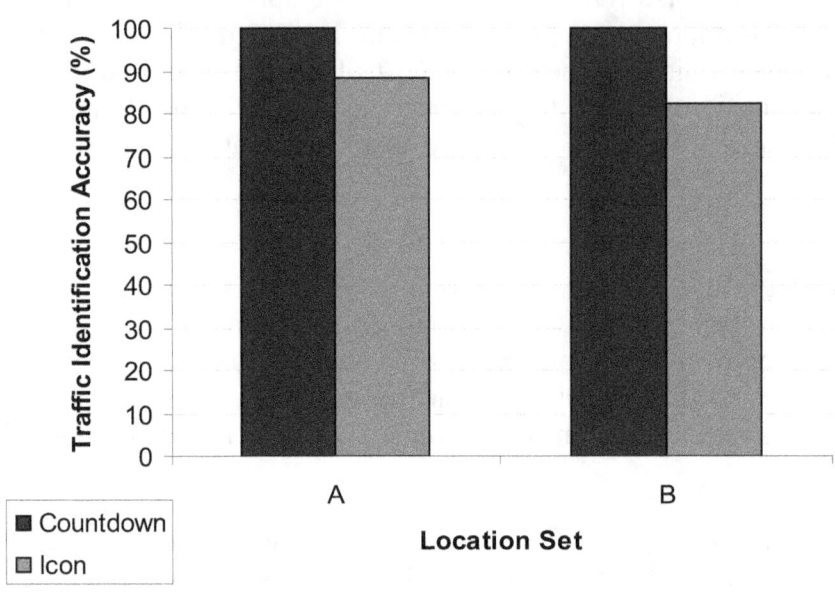

Figure 21. Accuracy of identifying traffic that the sign was giving information about for both location sets and genders.

3.2.1.3 Confidence in Identifying Traffic that the Sign ss Informing About

Participants were asked how confident they were in identifying what traffic the sign is telling information about. They did this on a five point scale of "Not at all confident," "somewhat not confident," "Neutral," "Somewhat Confident," to "Completely Confident" (see the "Trial 1" questionnaire pages in Appendix H). This measure was taken only during the participant's first exposure to each sign location/viewing location condition.

Pn. There were no significant differences in confidence between location sets when viewing them from Pn ($p = .278$; overall $M = 4.1$). There was a significant interaction between location and gender, $F(1,40) = 5.62, p = .023$. As shown in Figure 22, males felt more confident than females when viewing signs in location set A while the reverse was true while viewing location set B.

Pf. There were no significant differences in confidence between location sets when viewing them from the median ($p = .980$; overall $M = 4.5$).

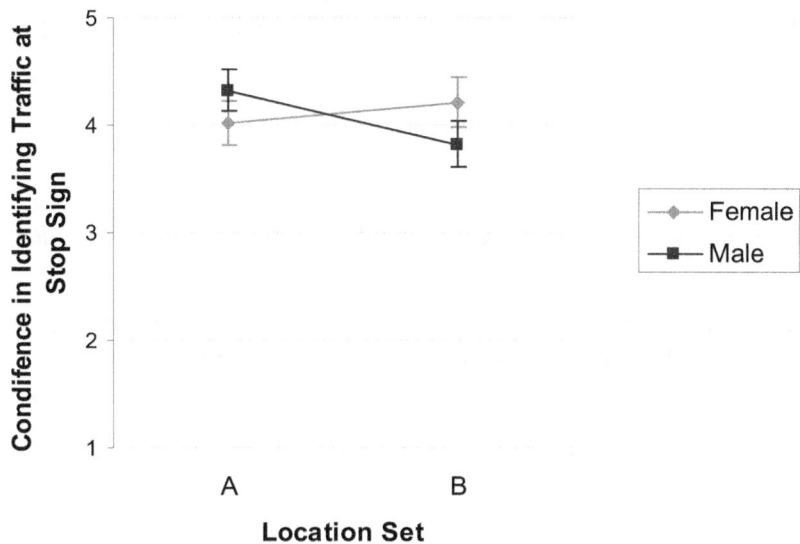

Figure 22. Confidence in identifying what traffic the signs were telling information about for both location sets and genders.

3.2.2 Comprehension Summary

There were no differences between the location sets in terms of reaction time for their first trial or for their first exposure to each location set/sign condition.

The accuracy results suggest that unpracticed observers performed well overall at accurately identifying what traffic the signs were telling information about regardless of location set or viewing location. When stopped at the stop sign, the A locations may also lead unpracticed sign observers to be accurate more often.

The confidence results suggest that unpracticed observers reported they were "somewhat confident" at accurately identifying what traffic both signs were telling information about regardless of location set.

3.2.3 Usability

For all of the usability questions, participants rated their agreement with statements on a five-point scale of "Strongly disagree," "Disagree," "Neutral," "Agree," to "Strongly Agree." Results from Pn at the stop sign (near) and Pf "median" (far) viewing locations were analyzed separately. Usability measures were collected after participants' second exposure to each sign location/viewing location condition (see the "Trial 2" pages in Appendix H).

3.2.3.1 It Was Easy to Associate Information on the Sign to Traffic Conditions

There were no significant differences in agreement between location sets when viewing them from Pn ($p = .454$; overall $M = 4.0$) or from Pf ($p = .801$; overall $M = 4.2$).

3.2.3.2 It Was Comfortable to View Sign in This Location

Pn. There were no significant differences in agreement between location sets when viewing them from Pn ($p = .270$; overall $M = 3.8$).

Pf. There was a significant difference between the location sets, $F(1,39) = 19.60$, $p < .001$, showing higher agreement for viewing comfort during location set B ($M = 4.4$) when compared to location set A ($M = 3.5$).

3.2.3.3 The Sign Obstructed My View of Approaching Traffic
Pn. There were no significant differences in agreement between location sets when viewing them from Pn ($p = .494$; overall $M = 1.8$). Participants reported higher agreement that the icon sign obstructed traffic ($M = 1.9$) when compared to viewing the countdown sign ($M = 1.6$), $F(1,37) = 8.96$, $p = .005$. The interaction between location set and sign type also approached significance, $F(1,37) = 4.00$, $p = .053$, suggesting that the icon sign was thought to obstruct traffic slightly more when in location set A ($M = 2.0$) than when in location set B ($M = 1.8$).

Pf. There was a significant difference between the location sets, $F(1,38) = 19.60$, $p < .001$, showing higher traffic obstruction during location set A ($M = 2.2$) when compared to location set B ($M = 1.5$). The interaction between location and age group approached significance, $F(1,38) = 3.87$, $p = .056$, suggesting that younger drivers may have thought the signs obstructed traffic more after viewing location set A ($M = 2.5$) than did older participants ($M = 2.0$) while all drivers thought location set B was less obstructing (both $M < 1.6$).

3.2.3.4 It Was Easy to See at this Distance

Pn. There were no significant difference in agreement between location sets when viewing them from Pn ($p = .067$; overall $M = 4.2$).

Pf. There was a significant difference between the location sets, $F(1,38) = 5.36$, $p = .026$, showing higher agreement for ease of seeing the sign during location set B ($M = 4.4$) when compared to location set A ($M = 3.9$).

3.2.3.5 Usability Summary

Unpracticed observers reported that the signs made it easy for them to associate information on the sign to traffic conditions regardless of location set.

They also reported that the location sets were equally comfortable to view when waiting at the stop sign. However, participants agreed that location set B was significantly more comfortable to view while viewing at the median.

Unpracticed observers reported that signs at both locations did not obstruct traffic when viewing signs from the Pn viewing location. While waiting at the median the sign in the median (set A) obstructed their view more than did the sign across the road (set B). They also thought the icon sign obstructed more traffic while at the stop sign (Pn).

Participants reported that the signs at both locations were easy to see at the distance observed when viewing signs from the stop sign. While waiting at the median the sign across the road (set B) was easier to see than the sign from the median (set A).

3.2.4 Comparison Preferences

After all conditions were completed, participants were asked their preference for location set A or B. Participants were also asked their preference for individual sign locations at Pn and Pf, for which they were given a labeled diagram of the sign locations (see Figure 23).

3.2.4.1 Layout Set Preference

When asked if they preferred layout set A or B (see Figure 20), there was not a statistically significant difference in preference ($X^2 = .818$, $p = .366$) although 57% of participants preferred location set A. When asked why they preferred their choice of locations, participants gave the responses shown in Table 22.

Table 22. The number of positive and negative responses to the question, "Why did you prefer this pair of locations?" split by location set.

	Location Set A	Location Set B
Why they preferred this location:	15 Easy to see traffic and signs	12 Easy to see traffic and signs
	…10 because I'm looking to left for traffic already; closer to my central field of view	…5 because does not obstruct view of traffic
	7 Easy to see signs	…4 because more natural to look across traffic; involves less head movements
	… 3 because it's closer	4 Easy to see signs
	2 Natural/familiar placement of sign; lets me know the sign is for me	…1 because comfortable, in line of vision
		2 Easy to see signs at median (Bf)
	1 Seems more comfortable	1 Easy to ignore if you want to use own judgment
Why they did NOT prefer this location:	3 Obstructs view of traffic	2 Blocked by car pillar
	3 May focus on sign, not traffic; distracting	2 Could be blocked by traffic
	2 May not be able to see if you pull too far forward	

Those that preferred set A reported that it was easy to see both the sign and traffic at the same time because it was near where they were looking anyway. Some said it was easier to see the signs because they were closer to their vehicle. Some that did not prefer set A thought this layout obstructed their view of traffic and that it may be distracting and would be easy to focus on the signs rather than the traffic itself.

Those that preferred set B reported that it was easy to see both the sign and traffic at the same time because the sign did not obstruct their view of traffic while also allowing them to make less scanning head movements. Similarly, some said it was easier to see the signs because it was comfortable to see them in their line of vision.

3.2.4.2 Sign Location Preference While Waiting at the Stop Sign

When asked their preference of sign location while they were at the stop sign (either An or Bn, as shown in the left portion of Figure 23), 63% of participants preferred the location immediately to their left (An) over the location across the near lanes in the median (Bn), although this was not a statistically significant difference ($X^2 = 2.81, p = .093$).

When asked why they preferred their choice of locations, participants gave the responses shown in Table 23. Those that preferred location An reported that it was easy to see both the sign and traffic at the same time because it was near where they were looking anyway. Some said it was easier to see the sign because they were closer to their vehicle.

Those that preferred location Bn reported that it was easy to see both the sign and traffic at the same time because the sign did not obstruct their view of traffic while also allowing them to make less scanning head movements.

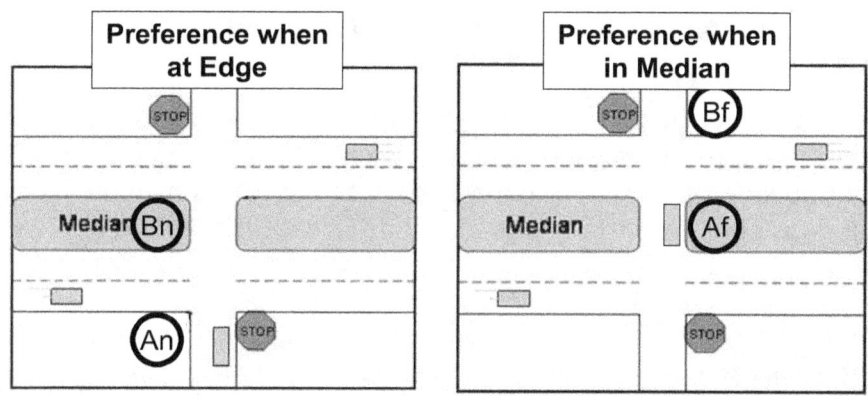

Figure 23. Preference choices when at Pn (An v. Bn) and Pf (Af v. Bf).

Table 23. The number of positive and negative responses to the question, "Why did you prefer this location?" for the near sign locations (An & Bn), split by location set.

	Location An	Location Bn
Why they preferred this location:	14 Easy to see traffic and sign	11 Easy to see traffic and sign
	…12 because I'm looking to left for traffic already; closer to my central field of view	…6 because does not obstruct view of traffic
		…2 because more natural to look across traffic; involves less head movements
	6 Easy to see sign	6 Easy to see sign
	…2 because it's closer	
	1 Natural/familiar placement of sign; lets me know the sign is for me	1 Wasn't too close, kept peripheral vision open
	1 Caught my attention quicker	
Why they did NOT prefer this location:	1 Obstructs view of traffic	1 Blocked by car pillar
	1 May focus on sign, not traffic; distracting	

3.2.4.3 Sign Location Preference While Waiting in the Median

When asked their preference of sign location while they were in the median (either An or Bn, as shown in the right portion of Figure 23), 58% of participants preferred the location across the far lanes on the opposite side of the intersection (Bf) over the location immediately to their right (Af), although this was not a statistically significant difference ($X^2 = 1.14$, $p = .286$).

When asked why they preferred their choice of locations, participants gave the responses shown in Table 24. Those that preferred location Af reported that it was easy to see the sign because they were closer to their vehicle. Some reported that it was easier to see both the sign and traffic at the same time because it was near where they were looking anyway.

Those that preferred location Bf reported that it was easy to see both the sign and traffic at the same time because the sign did not obstruct their view of traffic while also allowing them to make less scanning head movements. Some reported that it was easier to see the sign because it was further away and since it was in their forward line of vision it was more comfortable to view.

Table 24. The number of positive and negative responses to the question, "Why did you prefer this location?" for the far sign locations (Af & Bf), split by location set.

	Location Af	Location Bf
Why they preferred this location:	7 Easy to see sign …3 because it's closer 6 Easy to see traffic and sign …6 because I'm looking to left for traffic already; closer to my central field of view 1 Natural/familiar placement of sign 1 Caught my attention quicker	12 Easy to see traffic and sign …6 because does not obstruct view of traffic …3 because more natural to look across traffic; involves less head movements …1 because further away 10 Easy to see sign …3 because further away …2 because comfortable, in line of vision …1 because you should be looking forward 1 Caught my attention quicker
Why they did NOT prefer this location:	2 View of sign obstructed by rearview mirror and car pillar 1 May focus on sign, not traffic; distracting	

3.2.4.4 Comparison Preferences Summary

A small (but not statistically significant) preference was shown for layout set A with the strongest showing of support coming from participants in the older age group. Indeed, when asked what sign they preferred when at Pn, the results suggest that almost two-thirds of participants preferred the location from set A (An; although this was also not a statistically significant difference). Open-ended responses suggest that drivers liked how the signs at locations in layout set A were in the direction they were facing already, allowing them to easily view traffic and the sign. They also liked how they were close, making them easier to view. However, those who did not like location set A said they thought the signs obstructed their view of traffic and thought they might be distracting. Alternatively, those that preferred set B said that these locations were more comfortable and natural to view.

While at Pf participants had a small (but not statistically significant) preference for the location from set B (Bf), which appears to be due to the younger drivers overall support for location set B. Open-ended responses suggest that this change in preference is due to the signs at location Af being difficult to see and obstructing the respondents' view of oncoming traffic. Respondents also found signs at location Bf to be more natural and comfortable to view, mainly because they were further away and not obstructed by the rearview mirror or the car's pillar. Taken with respondents preference for location set A, this also suggests that signs at location Af might be improved if the sign was moved further from the median, such that their vehicle no longer interferes with viewing the sign and the sign does not obstruct their view of traffic.

3.3 Discussion

It was important to examine the location of the SSA signs because the information on the signs are intended to be associated with traffic patterns around the intersection. By testing location sets at the respective viewing locations, the consistency of the mental model between the intersection traffic and sign was examined in this study.

Accuracy of the participant's mental map was quantified in terms of their ability to identify which traffic the signs were displaying information about and the amount of time it took them to make this judgment. A summary of all results from the location study can be found in Table 25.

In terms of comprehending what traffic the signs were giving information about, location set A allowed drivers to be more accurate while they waited to cross from the stop sign. Participant's reaction time to the traffic identification task and confidence in their judgments was unaffected by where the signs were placed.

When asked how easy it was to associate information from the signs to traffic, drivers found no differences between the location sets. However, observers found location set B to be more comfortable to view, obstruct traffic less, and easy to see while waiting in the median. This also agrees with their post-study preference for the sign in location set B (Bf in Figure 20) while at Pf.

That said, drivers preferred the sign in location set A (An in Figure 20) while waiting at the stop sign. When asked if they preferred location set A or B, a simple majority of participants preferred set A. Respondents preferred viewing the signs in locations where they were already viewing (i.e., location set A), although they reported that the positioning of Af could be improved.

Table 25. Summary of the differences between location sets A and B.

Measures	Metrics	Location Set A	Location Set B
Comprehension	RT	*no difference*	
	Accuracy	Higher @ edge	
	Confidence	*no difference*	
Usability	Ease of Association	*no difference*	
	Viewing Comfort		Higher @ median
	Traffic Obstruction		Lower @ median
	Ease of Seeing		Higher @ median
Preferences	Location Set	57%	
	By View Location	63% @ edge	58% @ median

Note: "@ edge" and "@ median" indicate that the difference occurred only while at that particular viewing location.

3.4 Conclusions

The results suggest that observers took a relatively similar amount of time using signs at either location set and from either viewing location. Layout set A produced fewer errors in understanding which traffic the signs were referring to and this layout was preferred while viewing the signs from the stop sign. This last fact is important, in that some drivers may attempt to cross the intersection using a one-stage maneuver (without stopping in the median at P_f to reassess the situation). If set A is chosen it is also recommended that the Af location be modified. The current location proved troublesome to view since it was too close to the driver and was often reported as obscured by the participant's vehicle. The signs should be relocated such that the vehicle's A pillar or roof does not obstruct the driver's view of the sign, and also to make sure that the sign minimizes obstruction to the driver's view of traffic. On the other hand, respondents found location set B to be more comfortable to view, to obstruct traffic less, and to be easy to see while waiting in the median (P_f).

The overall preference scores between the two location sets were not significantly different, so it could be concluded that either location set choice could be implemented with adequate acceptance by drivers. However the lack of marked differences between location sets may also suggest the influence of factors unrelated to sign placement. In particular, we submit that there may be a need for more sensitive metrics to evaluate the relationship between sign placement and comprehension.

Based on our experience we also submit that examining potential limitations of the rotation and location of signs at an actual intersection would allow for the identification of factors that may have been implemented differently in the virtual environment. For example, a sign that is not obscured by the A pillar in a virtual environment may align slightly differently in the real world or may appear different to a driver seated at a different height from the road.

After the virtual testing described in this paper was completed, we found evidence to support this notion of differences between real and virtual environmental presentations. Wooden mock-ups of the approximate size of the SSA signs were placed at each of the four locations at the experimental intersection. Observations by the research team were made as to the visibility of the signs as well as how much they obscured expressway traffic from both stopping points (P_n and P_f). The consensus was that drivers, especially those seated in larger vehicles (e.g., heavy trucks) would have difficulty viewing oncoming traffic from both directions when signs were placed in location set A. This finding concurred with comments from testing in the virtual environment, even though those drivers were placed at a car's eye-height. In light of this finding, and because the project required a choice between the two originally proposed layouts, it was decided that layout set B was the preferential location set for the Random Gap study as well as future on-road implementations.

4 Random Gap Study

The optimal test of any SSA sign is how it may support driver performance and be usable for drivers in a real world environment. However, due to effort and financial limitations it is not possible to conduct an on-road study examining the utility of all three candidate signs (see Table 26). In light of this, the primary goal of the current study was to identify which of the three SSA candidate signs should be employed in field testing. The study evaluated the SSA interfaces in a simulated replication of the Minnesota test intersection (Hwy 52, CR 9) in order to maximize the generalization of results from the simulated test environment to the actual test intersection. Each of the three SSA candidate signs was compared against a baseline stop-sign only condition. To better understand how the use of each sign may influence driver performance metrics relative to accepted gaps, rejected gaps, safety margins, movement time, wait time, crossing maneuver type, and crashes were evaluated. In light of the notion that sign usability may impact significantly the employment of signs this study utilized subjective responses that included mental workload, usability, sign use, and sign preference.

Age and lighting conditions were also considered important to test during this study. Older drivers are over-represented in rural intersection collisions (Staplin & Lyles, 1991; Stamatiadis et al., 1991; Preusser et al., 1998) and may also have more difficulty understanding traffic signs and signals (Shinar et al., 2003; Dewar, Kline, & Swanson, 1994). Additionally, a 2002 safety audit and analysis of crash records for US 52 (including the test site) suggested there are more crashes in darkness than expected in comparison to similar rural highways (Preston & Rasmussen, 2002). In light of this finding it was also relevant to evaluate the proposed concepts under the suboptimal conditions represented by darkness when viewing conditions are limited and workload is expected to be higher. This is relevant in Minnesota given that morning and evening rush-hours often take place during dark hours in the winter.

4.1 Methods

4.1.1 Participants

Participants were recruited from the Twin Cities area with a specific effort to recruit drivers from the outer suburbs of the city, where they are more likely to encounter rural intersections resembling the test intersection. A second measure to ensure that participants were similar to drivers who typically drive through the test intersection included recruiting drivers that held a valid Minnesota drivers' license and reported driving at least occasionally in the past month. 60 participants completed the study. To examine the influence of age on driver performance and usability 30 participants (15 male and 15 females) were classified as Young (18-35 years of age) and 30 were classified as Senior (60+ years of age). Participants in each age group were randomly assigned to either a Day or Night driving condition in order to better understand the potential influence of time of day. This assignment process resulted in 15 participants per experimental condition (Age; Young, Senior and Time of Day; Day, Night). The Young group had a mean age of 25.2 ($SD=4.2$) while the Senior group had a mean age of 63.7 ($SD=3.2$).

4.1.2 SSA Interfaces

Design changes were made to the original SSA sign concepts based on the results of the comprehension study and from recommendations made by the project's Technical Advisory Panel (TAP) that were aimed at bringing the four candidate SSA interfaces in line with common information and colors used in the Manual on Uniform Traffic Control Devices (MUTCD) (see the Comprehension study results earlier in this technical report). These design changes resulted sign elements that changed to indicate the detection of a gap in traffic and the detection of a gap in traffic in which it is unsafe to proceed. Table 26 depicts each of the three candidate signs along with their changing elements.

4.1.3 Driving Simulator

The study was conducted using the HumanFIRST Program's driving environment simulator (Oktal; AutoSim) within the ITS Institute at the University of Minnesota. The driving environment simulator consisted of a full-sized Saturn vehicle with realistic operational controls and instrumentation, a high-resolution visual scene (1.96 arc minutes per pixel) projected to a 5-channel 210-degree forward field-of-view screen. The rear visual scene was projected onto a screen behind the driver and was visible in the vehicle's rear-view mirror. The side mirror views were provided by LCD panels placed on the side mirrors that presented a simulated side view of the driving environment. Auditory and haptic feedback were provided by a 3D surround audio system, subwoofer, car body vibration, and a three-axis electric motion system (roll, pitch, z-axis) system.

4.1.4 Simulated Test Intersection

To enhance the ability to identify behaviors and usability perceptions in the driving environment simulator that represent those that would be found at the actual test intersection (located at US Highway 52 and County State Aid Highway 9 in southern Minnesota) an exact replication of the test intersection was created in the simulator. Elements within the actual intersection (e.g., yield signs, median, test equipment, speed limit signs, lane locations, lane markings) were mapped using GPS and then included in the simulated environment at the same locations. To allow for a Time of Day comparison the simulator's night model presented a darkened scene which included headlight models for approaching traffic and a headlight model that illuminated the roadway in front of the participant's vehicle.

4.1.5 Warning Thresholds

Data has been collected at the test intersection since September 2004 that includes information about gap patterns, the size of gaps accepted by vehicle type (e.g., car, truck), the gaps rejected by drivers, and crashes that have occurred. Gorjestani et al. (2008) examined the patterns of rejected gaps at the intersection for maneuver type (right, left turns, straight crossing), time of day, vehicle type, the range of gap sizes available to the driver before making a maneuver, and time spent waiting at the intersection to help determine what the rejection threshold for the SSA should be. Overall, the results showed that the 80% gap rejection threshold was independent of time of day, vehicle type, time waiting and average available gap. This resulted in a rejected gap threshold (weighted average) of 6.5 s for crossing a set of lanes (two lanes) from the stop sign or from the median. This threshold was used as the baseline and 1 s was added to create the alert threshold to account for time it might take drivers to respond to a sign message before initiating a crossing maneuver. This 7.5 s alert threshold is in line with previous research at stop-controlled intersections that suggest 7.5 s as a minimum threshold for crossing (Harwood et al., 1999; Lerner et al., 1995) (i.e., the threshold at which it is unsafe to cross the intersection). This threshold applies to the near lanes when a driver is at the stop sign and to the far lanes when a driver is in the median.

Because research has also shown that drivers will almost always accept gaps greater than 12 seconds (Teply et al., 1997; Kittleson & Vandehey, 1991) information is only presented on the SSA when a vehicle is within 11 s of the intersection. This approach reduces the number of sensors required by the system at the intersection and will result in reduced installation and maintenance costs. The SSAs provide information about the near and far lanes when a driver is at the stop sign and information only about the far lanes when the driver is in the median. Therefore, a prohibitive message is shown for the far lanes when a vehicle is within the 11 s threshold *and* the driver is still at the stop sign. This is due to the notion that drivers may not always stop in the median (i.e., one stage crossing) and will need more than 11 s to cross the near lanes, median and far lanes. This prohibitive warning for the far lanes when a driver is at the stop sign is intended to encourage drivers to reassess the SSA information and the traffic when they reach the median thus promoting a two-stage crossing.

Because the warning algorithm is based on an 80% rejection threshold, it is assumed that approximately 80% of the rejected gaps will be smaller than the alert threshold when crossing each set of lanes. For drivers who regularly reject gaps smaller than the alert threshold, activation of the SSA will affirm their decision to reject a gap. For drivers who may want to accept gaps smaller than the threshold activation of the SSA warnings should capture their attention and potentially encourage them to reject unsafe gaps.

4.1.6 Randomized Traffic Streams

In an effort to create traffic streams that are representative of that at the actual intersection the simulator's traffic generation tool employed an algorithm based on the actual distribution and probability of gaps observed at the test intersection. The tool generated a unique pattern of traffic and gap sizes using the algorithm for each trial in the study. Traffic approached the intersection at 65 mph. The gap distribution generated in the simulator was similar to the actual distribution of gaps observed at the intersection.

4.1.7 Procedures

Participants first completed the informed consent process as mandated when conducting studies involving human subjects (see Appendix J). This was followed by a computerized demographic questionnaire that queried drivers about driving history and driving patterns (see Appendix A). An introduction to the study that included a generic description of how a dynamic SSA might work at the intersection was provided, but participants were not provided with specific operational and sign intent information about the interface designs (see Appendix K). Two 5-minute practice drives were conducted to familiarize participants with the simulator's operation and the simulated test intersection. There were three experimental sign conditions and a baseline condition. The baseline condition presented only a stop sign at the intersection. The SSA experimental sign conditions included the Hazard sign, the Countdown sign, and the Icon sign (see Table 26 for a depiction of each of the candidate signs). Participants performed three trials in each of the three experimental sign conditions with each block of three trials being counterbalanced across participants to eliminate potential order effects.

After completing each sign condition participants exited the vehicle to complete usability questionnaires Appendix N which also allowed them a short break (approximately five minutes) to prevent visual and physical fatigue. During this break participants completed the Modified Cooper-Harper (Wierwille & Casali, 1983; see Appendix L) to measure mental workload associated with the use of each of the four conditions. They then completed a Post-Drive Questionnaire (see Appendix M), indicated whether they employed the sign within that condition to help them make their crossing decisions, and were asked to explain why they did or did not use the SSA. After these questions participants received a description of the sign's messages and completed the usability questionnaire developed by Van der Lann, Heino, and de Waard (1997) to assess participants' perception of sign usefulness and acceptance of the sign (see Appendix N). Once a participant finished all four sign conditions they ranked the three SSA signs based on their preference and the usefulness of the sign in making crossing decisions (see Appendix O).

Table 26. CICAS-SSA interfaces. Each interface displays multiple messages depending on whether the driver is at the stop sign or in the median.

Intended Meaning	Driver at Stop Sign		Driver in Median	
	Alert Condition	Sign Message	Alert Condition	Sign Message
Not Safe to Enter Lanes	Traffic detected within threshold of near or far lanes		Traffic detected within alert threshold in far lanes	

		Intersection Sign	Median Sign
Proceed with Caution	No traffic detected within threshold		No traffic detected within threshold in far lanes	
Not Safe to Enter Lanes	Traffic detected within threshold of near lanes		Traffic detected within threshold of far lanes	
Do not cross or turn into far lanes; may be ok to enter near lanes	Traffic detected outside threshold in near lanes. Traffic detected within threshold in far lanes.		Not applicable	Not applicable
Proceed with Caution	No traffic detected within thresholds for near and far lanes		No traffic detected within threshold for far lanes	
Do Not Enter	Traffic detected within threshold for near and far lanes		Traffic detected within threshold for far lanes*	
Do not cross or turn into far lanes; may be ok to enter near lanes	Traffic detected outside threshold in near lanes. Traffic detected within threshold in far lanes.		Traffic detected outside threshold in far lane	
Proceed with Caution	No traffic detected within threshold for near and far lanes		No traffic detected within threshold for far lanes	

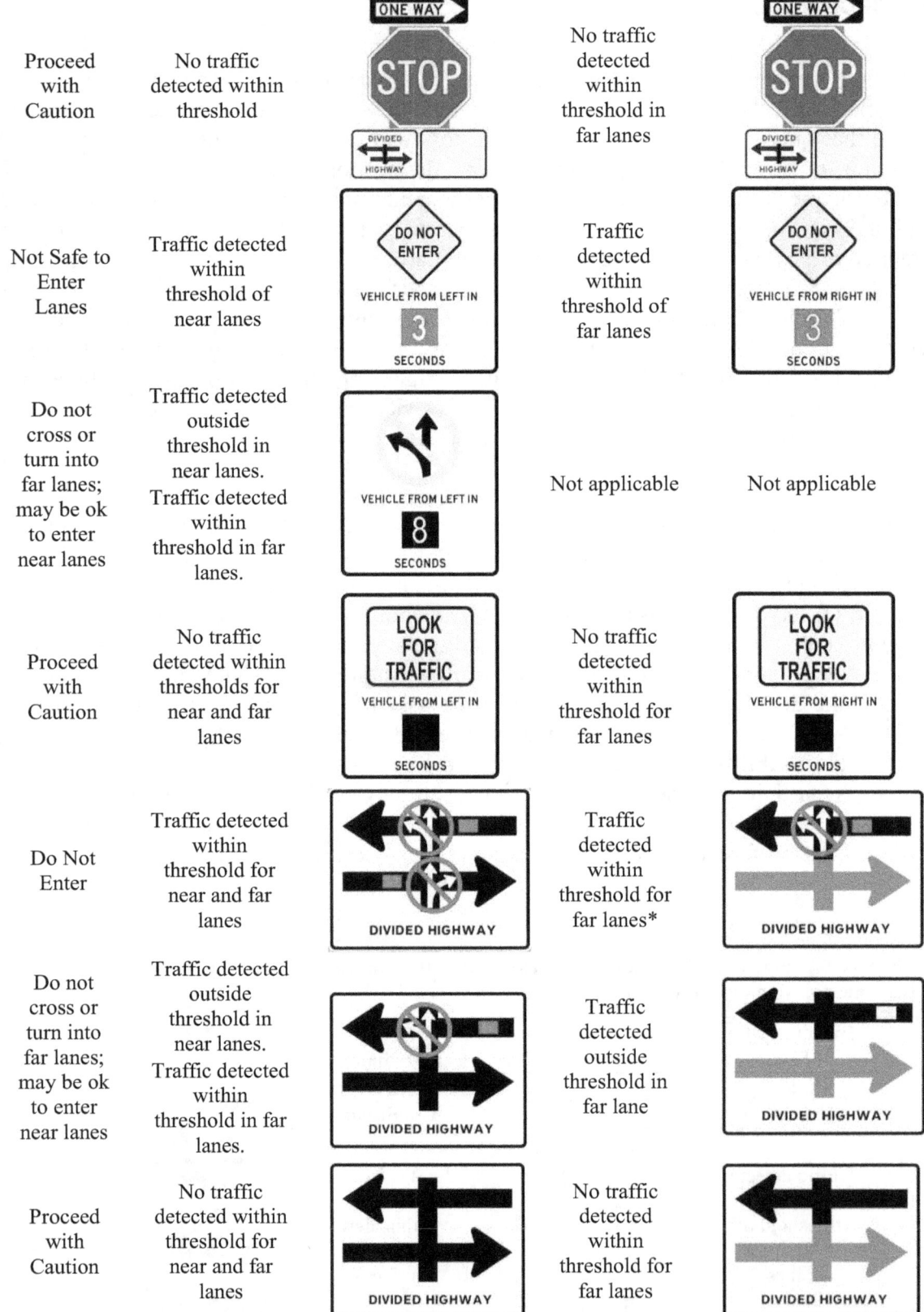

* Note: The bottom portion of the median sign continues to show icons for near lanes, however, the bottom is dimmed to draw attention to the top of the sign.

4.1.8 Statistics

The independent variables included in this study were Sign (Baseline, Hazard, Countdown, Icon), Age (Young, Senior), and Time of Day (Day, Night). Dependent variables were categorized according to performance and usability constructs.

Performance Construct Dependent Variables - The performance construct consists of those dependent variables that provide an indication of driver behavior changes relative to the employment of each of the candidate signs. The following list presents the performance construct dependent variables along with their definition.

- Accepted Gaps - The size of the accepted gap taken by drivers includes the total length of the available gap from the rear bumper of the lead vehicle to the front bumper of the following vehicle (see Figure 24). This metric was calculated for both Near Lane Accepted Gaps and Far Lane Accepted Gaps.
- Rejected Gaps - The distribution of rejected gaps across trials and sign conditions was examined in comparison to the rejected gap distributions of data from the intersection.
- Safety Margins – The safety margin is the time-to-contact (TTC) between the approaching vehicle on the major road and the participant's vehicle when it is the middle of the approaching vehicle's lane while crossing. It was measured for the near and far lanes. This metric was calculated for both Near Lanes Safety Margins and Far Lanes Safety Margins.
- Movement Time - Movement time is the total time to cross each set of lanes (near and far) calculated from when the front bumper of the participant's vehicle enters the first lane of the set to when the back bumper exits the second lane in the set. Slower movement times across a set of lanes could reduce the safety margins for a slower driver. This metric was calculated for both Near Lanes Movement Time and Far Lanes Movement Time.
- Wait Times - Wait time is the amount of time spent waiting at either the stop sign or in the median before crossing. Wait time is dependent on the gaps available to drivers over time when an SSA is not present (i.e., baseline condition). It is also possible that the presence of the SSA may increase wait time at the intersection. A poor design might result in longer wait times as drivers attempt to comprehend the sign's messages and use it. Alternatively, wait times could increase with the SSA because drivers are encouraged to reject a series of unsafe gaps in favor of waiting for a more acceptable gap. It is expected that a good SSA design may increase wait time as a function of safety but not excessively when compared to baseline. Wait time was measured when drivers were at the Stop Sign and in the Median. This metric was calculated for both Wait Times at the Stop Sign and Wait Times at the Median.
- Crossing Type Maneuver - Drivers can make either a one-stage crossing maneuver, which means they do not stop in the median before entering the far lanes of traffic, or they can make a two-stage crossing maneuver, which means they stop in the median before entering the far lanes. A two-stage maneuver is considered safer because drivers who stop have the opportunity to take more time to assess the traffic in the far lanes. At the Minnesota test intersection most crashes occur in the far lanes (Preston et al., 2004).

Evaluations of stop-controlled intersections in partner states, such as Wisconsin and Iowa, also found several intersections with significantly more far-lane crashes than near-lane crashes (e.g., Preston, Storm, Donath & Shankwitz, 2006; Preston, Storm, Donath & Shankwitz, 2007). Therefore, a goal of the SSA is to encourage drivers to make a two-stage maneuver by providing information at the stop sign and at the median. It was expected that more two-stage maneuvers would occur in the SSA sign conditions compared to Baseline.
- Crashes – A crash was recorded if a participant's vehicle and any vehicle in the driving scenario contacted each other.

Usability Construct Dependent Variables - The usability construct consists of those dependent variables that provide an indication of driver subjective perceptions of workload, usability, sign use and preference relative to each sign. The following list presents the usability construct dependent variables along with their definition.

- Rating Scale Mental Effort – The RSME is a univariate scale for rating mental effort (Zijlstra, 1993). It was presented on paper as a single continuum with specific points marked with workload descriptions. Operators marked the place on the continuum that best described the level of workload generated to interact with the SSA. Higher ratings indicate that greater extended effort was given to the task (see Appendix P for a copy of the RSME).
- Usability Scales – The Usability Scales are a measure of usability in terms of the perceived satisfaction and usefulness of the system (as described in Van der Laan, Heino & de Waard, 1997). This measure requires subjects to rate their perceptions on a number of bipolar adjective scales. These scales are then summed to produce separate scores for the level of perceived satisfaction and usefulness. These scores can be positive or negative with larger values representing greater satisfaction and usefulness (see Appendix N for a copy of the Usability Scales).
- Post Drive Questionnaire – This measure was a 10-item questionnaire where they rated their perceptions of each sign on a 5-point Likert scale (1 = strongly disagree; 5 = strongly agree) covering dimensions such as likeability, trust, and confidence in the information (see Appendix M).
- Sign Use – The sign use measures asked participants whether they employed the sign within a condition to help them make their crossing decisions and why they did or did not use the SSA (the sign use question appears in Appendix M).
- Sign Preference – Participants were asked to rank the signs based on an assessment of their personal preference for a design and how useful they felt the design was for supporting crossing decisions at the intersection. A rank of "1" was the most preferred and a rank of "3" was least preferred (the sign preference question appears in Appendix O).

A 4 (Sign: Baseline, Hazard, Countdown, Icon) x 2 (Age: Young, Old) x 2 (Time of Day: Day, Night) mixed-model analysis of variance (ANOVA) was performed for each of the dependent variables. Less than 1.5% of trials were missing data for each variable; therefore, no adjustments were made for the analysis regarding missing data. Before the analysis was conducted, the trial data was examined to determine if learning effects occurred within the sign conditions; a situation that would confound the study results. The analysis indicated no statistically significant differences across trials for the size of accepted gaps, time-to-contact, safety margins, or wait times (p's > 0.05). Therefore, trial values were averaged for each participant within a condition and these values were used in the overall analysis. Reported p-values reflect the Greenhouse-Geisser adjustment for sphericity. Differneces between means were considered significant at the $p<.05$ level. A Bonferonni correction (p<.05) was used to evaluate p-values for post-hoc tests. Only significant main effects or interactions are presented.

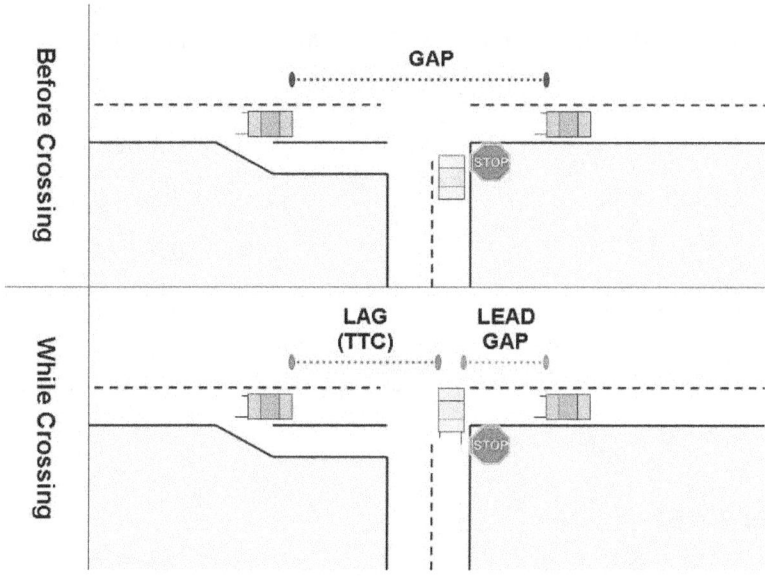

Figure 24. Visual description of "gap", "lag" and "lead gap".

4.2 Results - Driving Performance

Only significant main effects or interactions are presented.

4.2.1 Accepted Gaps

Far Lanes Accepted Gaps - There was a statistically significant effect of light condition on the far lane mean accepted gaps, $F(1,56)=8.49$, $p=0.005$, where the Day condition had a larger mean accepted gap ($M=7.96$ s) than the Night condition ($M=7.23$ s).

4.2.2 Rejected Gaps

On average, participants rejected 80% of all gaps that were smaller than 7.5 s across conditions for the near lanes and more than 90% of all gaps that were smaller than 7.5s for the far lanes (see Table 27. Percentage of gaps rejected that were smaller than 7.5 s. Overall, the Countdown and the Icon sign conditions showed the largest percentage of rejected gaps below the alert threshold for both sets of lanes.

Table 27. Percentage of gaps rejected that were smaller than 7.5 s.

	Near Lanes	Far Lanes
Baseline	81.7%	90.8%
Hazard	79.3%	93.0%
Countdown	83.0%	93.2%
Icon	84.5%	92.7%

4.2.3 Safety Margins

Near Lanes Safety Margins - There was a statistically significant ME of sign condition for the near lanes safety margin, $F(3,144)=3.53$, $p=0.02$ (see Figure 25). The Countdown sign condition had a significantly smaller safety margin ($M=6.50$ s) than the Baseline condition ($M=7.41$ s), $t(57)=3.78$, $p<0.001$. There was also a statistically significant ME of Age, $F(1, 48)=7.98$, $p=0.007$, where Old participants (M=7.22 s) had a larger average safety margin than Young participants (M=6.55 s). A significant ME of Light condition also existed, $F(1, 48)=11.07$, $p=0.002$, where, on average, the safety margin was higher in the Day ($M=7.28$ s) compared to the Night ($M=6.49$ s) driving condition. There was also a statistically significant interaction of Age by Time of Day condition, $F(1, 48)=4.46$, $p=0.04$ but none of the follow-up comparisons were significant.

Far Lanes Safety Margins - There was a marginally significant main effect of sign condition for the far lanes safety margin, $F(3,144)=2.74$, $p=0.05$; however, none of the post-hoc comparisons were significant. There was also a significant difference in safety margins for the Time of Day condition, $F(1, 48)=8.44$, $p=0.006$, where the Day ($M=7.48$ s) condition exhibited a larger safety margin compared with the Night ($M=6.75$ s) condition. See Figure 26 for a depiction of the main effect.

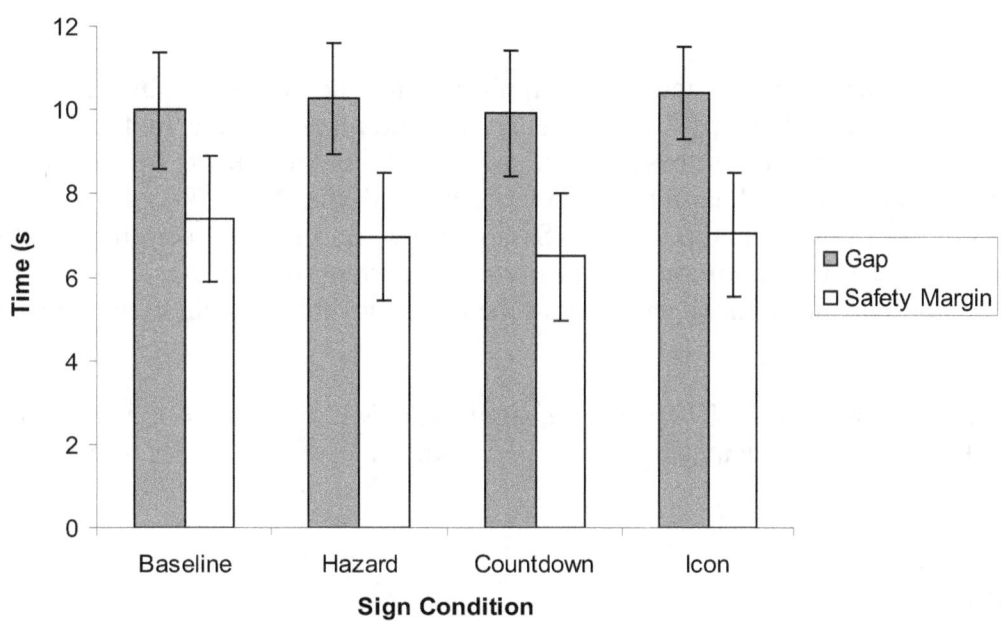

Figure 25. Means and standard deviations of accepted gaps and safety margins by sign condition for the near lanes when crossing from the stop sign.

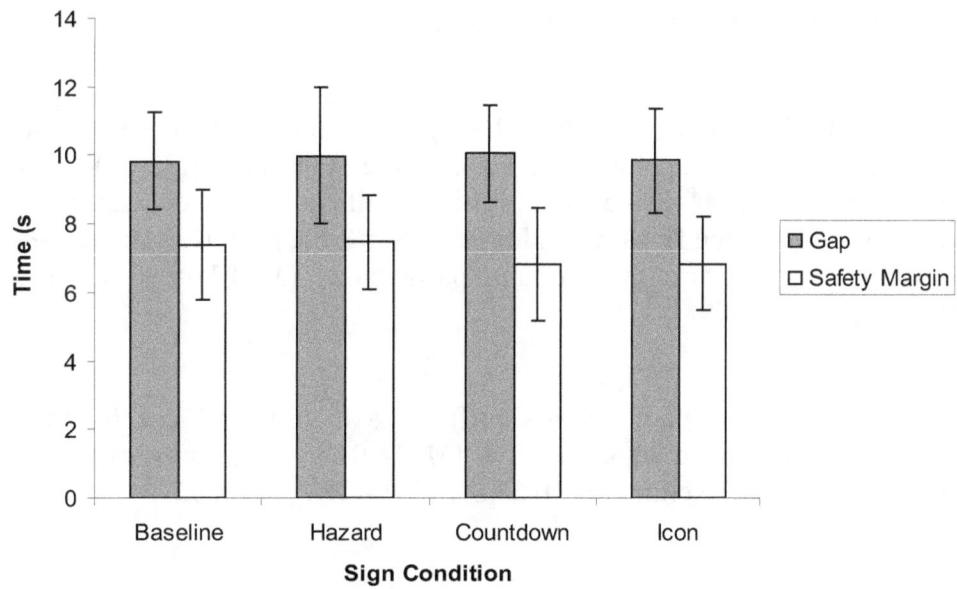

Figure 26. Means and standard deviations of accepted gaps and safety margins by sign condition for the far lanes when crossing from the median.

4.2.4 Movement Time

Near Lanes Movement Time - There was a significant interaction of Age by Light condition for the near lanes, $F(1,52)=4.30$, $p=0.043$, where the Senior participants in the Night condition (M=3.74 s) took longer to cross these lanes than the Young participants for Day (M=2.67 s) or Night (M=2.92 s) and the Old participants in the Day condition (M=2.79 s). This result suggests that the shorter mean safety margin for the Senior drivers in the night condition compared with Senior drivers in the day condition could be due, in part, to this slower crossing time, which would let the approaching vehicle get closer to the crossing vehicle while in the intersection.

Far Lanes Movement Time - There was a significant ME of Age condition, $F(1, 52)=9.05$, $p=0.004$, where Senior participants (M=2.23 s) had, on average, slower crossing times than Young participants (M=2.01 s).

4.2.5 Wait Time

Wait Times at the Stop Sign - There was a significant ME of Wait Time at the stop sign for the sign condition, $F(3,168)=17.7$, $p<0.001$. On average, the Hazard sign ($M=43.31$ s) had the longest wait time compared to Baseline ($M=18.14$ s; $t(59)=-4.78$, $p<0.001$, to the Countdown condition ($M=23.86$ s; $t(59)=3.81$, $p<0.001$, and to the Icon condition ($M=23.15$ s; $t(59)4.37$, $p<0.001$) (see Figure 27). The Icon sign also had a significantly longer wait time than Baseline at the stop sign, $t(59)=-3.16$, $p=0.003$ while the Countdown sign had a marginally significantly longer average wait time than Baseline, $t(59)=-2.72$, $p=0.009$.

There was a significant ME of Age, $F(1,56)=9.39$, $p=0.03$; however there was also a significant interaction of Sign by Age group for Wait Time at the stop sign, $F(3,168)=3.80$, $p=0.04$. On average, Senior participants ($M=57.45$ s) exhibited significantly longer wait times than Young participants ($M=29.17$ s) for the Hazard condition, $t(58)=-2.65$, $p=0.01$. Seniors ($M=30.07$ s) also had longer average Wait Times than Young participants ($M=17.66$ s) for the Countdown condition, $t(58)=-2.72$, $p=0.009$.

Wait Times in the Median - There was a significant ME of Wait Time for the median, $F(3, 168)=2.99$, $p=0.036$. The Hazard sign condition ($M=14.01$ s) had, on average, a longer wait in the median compared to the Baseline ($M=10.53$ s), $t(59)=-2.92$, $p=0.005$.

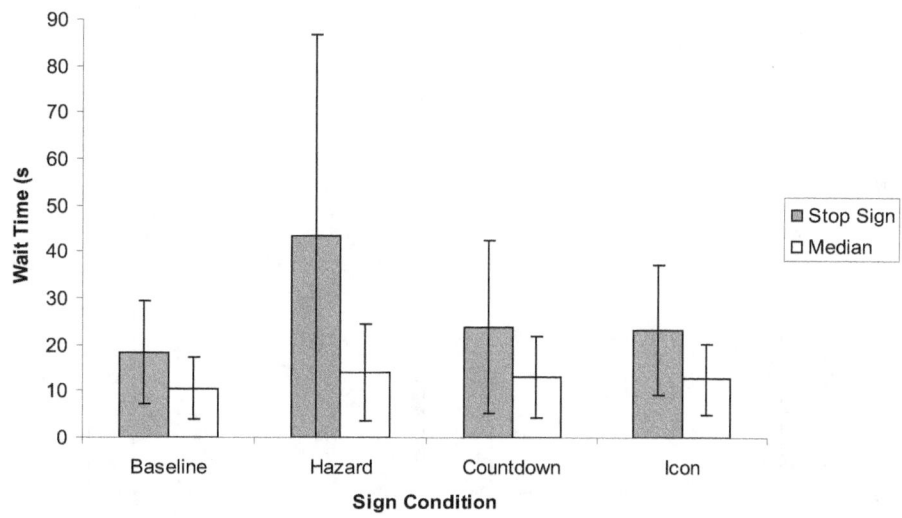

Figure 27. Means and standard deviations of Wait Time for waiting at the stop sign and in the median.

4.2.6 Crossing Maneuver Type

Across trials for each sign condition, most participants made two-stage maneuvers. Figure 28 presents the percentage of participants who made one-stage maneuvers during each trial for each sign condition. Overall, the most one-stage maneuvers occurred in the Countdown condition. An evaluation of maneuver type by safety margin across trials showed that one-stage maneuvers had smaller average safety margins overall when compared with two-stage maneuvers for both the near and far lane crossings.

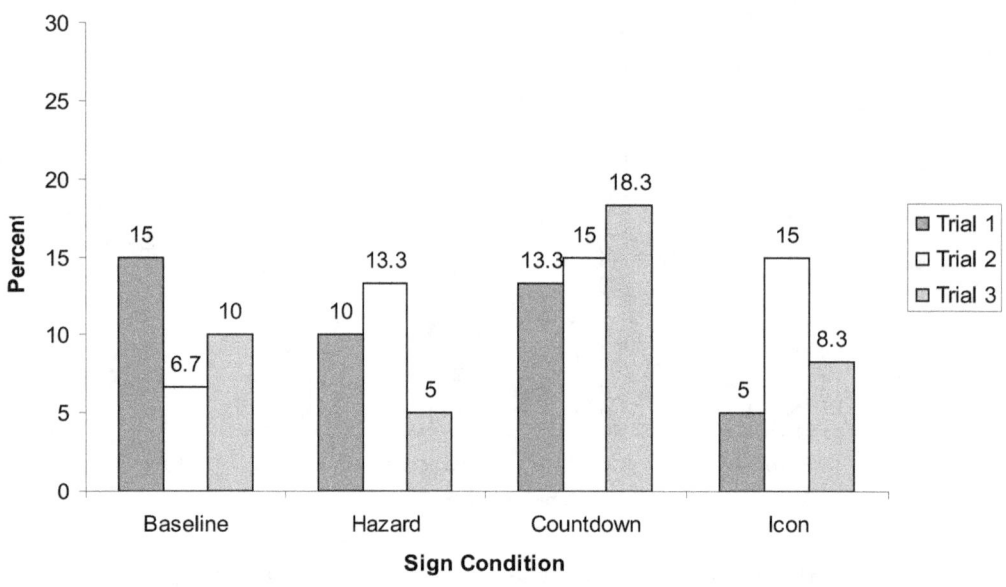

Figure 28. Percentage of participants who made one-stage maneuvers per trial for each sign condition.

4.2.7 Crashes

Six simulated crashes occurred during the study and all occurred during the Night driving condition. Five of the crashes occurred with the senior drivers. Three crashes occurred in the near lanes and three in the far lanes, and crashes occurred in all four sign conditions (1 Baseline; 2 Hazard; 2 Countdown; 1 Icon).

4.3 Results - Subjective

4.3.1 Post-Drive Questionnaire

Eight questions were statistically significant (p's<0.05) and in each question the Hazard sign was rated less favorably than the Icon or Countdown sign (see Appendix M). There were no statistically significant differences in ratings between the Countdown and Icon signs for any of the questions. For questions 1 and 3-9, higher agreement indicated a more favorable rating for questions 1 and 3-9, while higher agreement indicated a less favorable rating for questions 2 and 10.

Table 28. Results for the 10-item post-drive questionnaire.

Question		Hazard	Countdown	Icon	ANOVA Results
1	I felt confident using this sign.	2.50	3.50*	3.22*	$F(2,110)=11.30, p<0.001$
2	I felt it was confusing to use this sign.				$p>0.05$
3	Using this sign made me feel safer.	2.53	3.22*	3.10*	$F(2,110)=5.86, p=0.004$
4	I trusted the information provided by the sign.	2.58	3.78*	3.53*	$F(2,110)=20.88, p<0.001$
5	I like this sign.	2.52	3.12*	3.14*	$F(2,110)=4.04, p=0.021$
6	The sign was reliable.	2.70	3.85*	3.63*	$F(2,110)=20.86, p<0.001$
7	I felt this sign was easy to understand.				$p>0.05$
8	The sign's information was believable (credible).	2.82	3.87*	3.70*	$F(2,110)=18.2, p<0.001$
9	This sign was useful.	2.57	3.62*	3.35*	$F(2,110)=11.98, p<0.001$
10	I could complete the maneuver the same way without using the sign.	4.33	3.95*	3.98*	$F(2,110)=4.19, p=0.023$

* Indicates result was statistically significant when compared to the Hazard sign condition.

4.3.2 Usability Scales

Both the Countdown and Icon sign were rated somewhat useful and satisfying while the Hazard sign was rated slightly useful but not satisfying (see Figure 29).

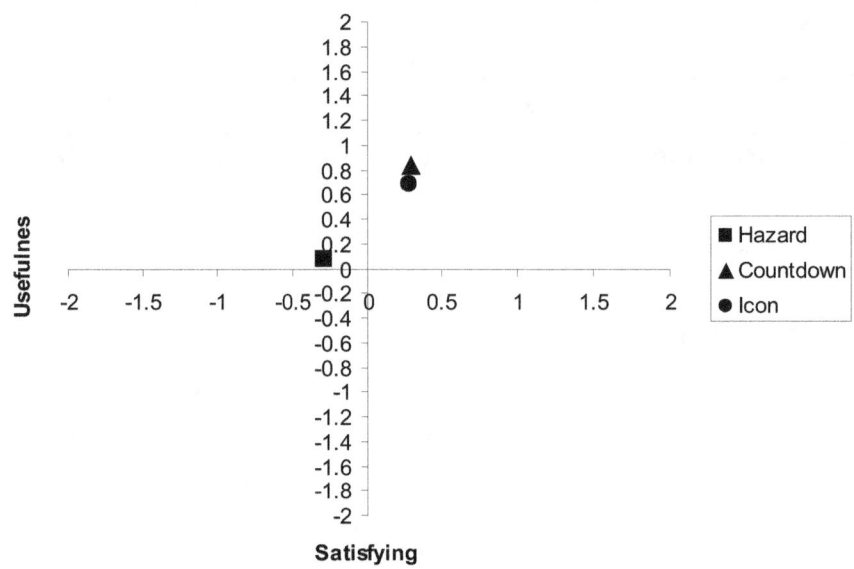

Figure 29. Usefulness and satisfying ratings for each sign.

4.3.3 Sign Use

Only 30.5% of participants said they used the Hazard sign to help them make their crossing decision compared with 81.4% who said they used the Countdown sign and 67.8% who said they used the Icon sign. Comments related to the Hazard sign indicated participants did not fully understand how the Hazard sign functioned at the stop sign.

4.3.4 Sign Preference

A rank of "1" was the most preferred and a rank of "3" was least preferred. The Wilcoxon's signed ranks test was used to evaluate the rankings. The Countdown sign was significantly more preferred (M=1.77) than the Hazard sign (M=2.35; p=0.001).

4.4 Discussion

This study evaluated driving performance and usability for three candidate SSA sign designs compared to a baseline condition. The goal was to identify the final candidate sign to be field tested at the Minnesota test intersection. Overall, the results showed that gap acceptance and gap rejection behavior was not significantly different between the SSA conditions and the baseline crossing condition. Performance in the simulator mirrored real-world driver behavior data from the test intersection. Mean gap acceptance overall in the simulator was approximately 10 s compared with 9.4 s (for straight across maneuvers) at the test intersection (Alexander et al., 2005). Participants in the simulator study rejected more than 80% of gaps that were smaller than the alert threshold, which is similar to the gap rejection pattern observed at the real intersection. The similarity in behavior between the SSA conditions and the baseline conditions in the simulator for these two measures is expected because the alert threshold is derived from actual gap rejection behavior at the real intersection. The assumption for using the 80% rejection threshold as the basis for the alert threshold is that drivers are generally good at rejecting unsafe gaps at intersections (Gorjestani et al., 2008). Therefore, analysis of other safety-related behaviors must be taken into account to determine the safety of performance when using the SSA signs.

One example of an unsafe behavior that can lead to a crash is a one-stage crossing maneuver where a driver fails to yield in the median and re-assess the oncoming traffic arriving from the right. Research at rural stop-controlled intersections during the IDS project indicated that the majority of crashes occur in the far lanes, regardless of geometry or sight lines (Preston et al., 2004; 2005; 2007). This indicates that drivers have a problem assessing the far-lane gap from the stop sign and that they frequently fail to further assess the far-lane gap when in the median. The design of the SSA supports this known crash risk by including information about the far lanes at the stop sign and again in the median to encourage drivers to re-evaluate gaps in the far lanes. In this study, an examination of safety margins by maneuver type revealed that one-stage maneuvers had smaller average safety margins compared with two-stage maneuvers for both sets of lanes. This suggests drivers who make one-stage maneuvers may be riskier overall as they tend to have smaller safety margins in both sets of lanes, not just the far lanes. The SSA may assist these drivers in learning a safer threshold for crossing.

The Icon sign and Hazard sign conditions both had similar rates of two-stage maneuvers compared to baseline whereas the Countdown sign had the highest rate of one-stage maneuvers. The Countdown sign design may have encouraged one-stage maneuvers rather than discouraging them in the simulator study and, in fact, participants reported calibrating their own judgments to the timer in the Countdown condition, regardless of whether it was above or below the alert threshold. For example, one participant wrote, "Anytime it was a 5+ second count I felt I had enough time to cross". This desire to "beat the clock" or to calibrate the Countdown timer to one's current behavior was also observed in the original IDS study with this sign's concept design (Creaser et al., 2007). The replication of this result indicates the Countdown sign is not a suitable candidate for the final SSA design.

Another problem with gap acceptance at intersections occurs when drivers are in a hurry or become impatient while waiting for an acceptable gap (Caird & Hancock, 2002). In this case, drivers may accept smaller gaps than they normally would in order to traverse the intersection more quickly (Pollatschek, Polus, & Livneh, 2002). The SSA can potentially encourage drivers to wait for a larger gap. On average, participants waited longer at the stop sign before taking a gap in the SSA conditions when compared to baseline. For the Countdown and Icon signs, this difference was 5 s greater or approximately one gap length longer. Increased wait times at the intersection for SSA conditions may also indicate a shift towards safer gap acceptance behavior. This suggests a small shift in behavior towards accepting larger gaps during the SSA conditions. Because traffic streams were random for each trial, the longer wait times in the Countdown and

Icon conditions suggest participants were responding to the information on the signs and may have delayed gap acceptance based on SSA information.

However, excessive waiting due to SSA use could also be problematic. The Icon sign did not show excessive waits times. In contrast, wait time was excessively increased in the Hazard condition at the stop sign and a review of sign use and comments showed participants did not understand the flashing "Traffic Too Close" message when at the stop sign. Participant comments suggested confusion related to the Hazard sign's continuous flashing when traffic was detected in the far lanes but not the near lanes. This indicates that a simple design for displaying alert and gap information about the near and far lanes simultaneously was not effective. The lack of excessive wait time for the Hazard sign in the median when it only provides information about one set of lanes supports this conclusion. The Countdown sign showed an age effect for wait time where older drivers waited significantly longer than younger drivers. Older drivers may have had a more difficult time comprehending the sign, or the younger drivers may have been more likely to calibrate themselves to a shorter gap and, thus, entered the intersection sooner than the older participants. The results for both the Hazard and Countdown signs suggest neither were optimal for displaying gap information to drivers at the intersection.

Another concern when designing the SSA signs was how information processing might be affected. Because the SSA is a decision support system, it was important that comprehension and responses to the sign's information not affect drivers' abilities to act quickly to enter the intersection once a gap decision is made. Safety margins are most likely to be affected by processing demands as a decrease in safety margin would occur for drivers who are slow to react to the information on the sign, even if the original gap is sufficient for crossing. There were no differences in safety margins for either set of lanes when comparing the Hazard and Icon signs to the baseline condition. Because so few participants reported using the Hazard sign to help them with their crossing decisions, it is expected that safety margins are similar given that the accepted gaps were similar for the two conditions. In comparison, 67.8% of the participants reported using the Icon sign to help them make their crossing decisions. Because the mean accepted gap was also similar for the Icon and baseline conditions, a lack of difference in safety margins is also expected. However, the Icon condition is not equivalent to the baseline condition because it requires drivers to view and comprehend information on the sign. A lack of difference in safety margins compared to baseline indicates Icon sign use did not delay drivers from entering the intersection after an appropriate gap was identified.

In the Countdown condition, 81.4% of participants said they used the sign to help them make their decision. The Countdown sign had an approximately 1 s smaller safety margin than the baseline conditions for the near lanes. This difference in safety margin is not accounted for by differences in the mean accepted gap or by differences in movement time across the intersection between the two conditions. This indicates participants were slower to enter the intersection after making their gap acceptance decision in the Countdown condition. Because participants did not rate the Countdown sign as requiring more mental effort than the Icon sign to comprehend, design differences may not be the sole cause of the reduced safety margins in the Countdown condition. Instead, a significant effect of processing time may have been accentuated in the

Countdown condition because more participants reported using the Countdown sign than the Icon sign. This explanation is supported by the lack of a significant difference in safety margins between the Countdown and Icon sign conditions.

In addition to driver performance, a number of usability measures were also collected to evaluate the signs subjectively. The Hazard sign was least preferred by participants along a number of dimensions (e.g., reliability, trust, usefulness, perceptions of safety) but no differences in preference existed between the Icon and Countdown signs. The Hazard sign was also rated the least useful and satisfying of the three SSA designs while the Countdown and Icon signs were rated similarly as somewhat useful and satisfying. Participant comments indicated that the design of the Hazard affected their perceptions of it. Drivers were not aware and could not easily figure out that the sign alerted for traffic in both sets of lanes when they were at the stop sign.

4.4.1 Age

Differences in wait time between young and senior drivers for the Hazard and Countdown signs was the only age effect seen in this study related to SSA performance. However, several results related to differences between young and senior drivers were observed. For example, senior participants had larger average safety margins in the day driving condition compared to Young participants; however, their gap acceptance behavior was similar. Senior participants also had slower movement times across the near and far lanes compared to Young drivers in the night condition and senior participants in the day condition. This partially explains the difference in safety margins between the Senior groups (day and night), but not the lack of difference between senior and young drivers at night, which had similar safety margins. Slower movement times were also seen in the first IDS concept study (Creaser et al., 2007) and have been reported in other research as well (Hakamies-Blomqvist, 1996; Keskinen, Ota, & Katila, 1998; Lerner et al., 1995). Senior drivers also report that night driving is more difficult for them than day driving (e.g., Holland & Rabbit, 1992; Benekohal, Michaels, Shim, & Resende, 1994). If senior drivers in this study found the night driving to be more difficult, they may have been more cautious with their entries into the intersection once selecting a gap. However, slower movement times across the intersection may also represent a hazard for senior drivers. Five of the six crashes were for senior drivers crossing in the night condition.

4.4.2 Light Conditions

The use of the night driving condition allowed an examination of driver behavior with the SSA signs under non-optimal visibility conditions. Overall, there were no significant interactions of light condition by sign condition. In general, results showed slightly smaller mean accepted gaps and mean safety margins (about ¾ of a second) for night versus day driving. This may have resulted from drivers' differing ability to detect oncoming vehicles between the night and daytime conditions, where the daytime condition presented drivers with more visual cues for detecting the speed and location of oncoming vehicles.

4.5 Limitations

Although the simulator provided an exact replication of the real-world test intersection and provided gap streams similar to what are observed in the real-world there are limitations with the simulator study. First, this study only examined straight crossing maneuvers and did not examine cases with drivers turning left or right. This decision was made because crash data from the intersection showed most crashes occurred for drivers crossing the intersection rather than turning. However, because the sign is intended to support all crossing decisions at the intersection, the field test will examine performance for all maneuver types. Additionally, field operational testing is recommended to determine if driver behavior changes at the test intersection after installation of the signs. Second, even in the experimental setting, the SSA signs did not produce full compliance with the recommended thresholds. This may have been due to the nature of the study and its requirement that participants not be told what the signs mean before interacting with them.

4.6 Conclusions

The results of this study when compared to data collected at the test intersection indicate that the driving simulator can be used to replicate real-world geometries and traffic flows for testing the CICAS-SSA signs. Testing the final design this way reduces the risk involved in future field testing by narrowing down the best sign design before implementation in the real world. This study indicated that, similar to the real world, drivers in the simulator were good at rejecting unsafe gaps at rural stop-controlled intersections in both the SSA and baseline conditions. However, the presence of the SSA affirms good decision making while also supporting drivers who may have difficulty in the selection and acceptance of a safe gap when crossing. For the Countdown and Icon sign conditions, drivers reported using the SSA information to help with their crossing decisions. However, performance data showed that the Countdown sign also resulted in behaviors that may elevate the risk of a crash, such as one-stage crossing maneuvers and misuse of the timer functionality. Although a similar distribution of unsafe gaps were rejected while using the Icon sign, participants chose gaps with larger safety margins and exhibited more two-stage crossing maneuvers with this sign compared to the Countdown condition. Overall, the Icon sign resulted in performance similar to the baseline condition and was ranked equally well on the usability measures when compared to the Countdown sign. It is recommended that this sign be implemented in an experimental field test. The field test will use similar measures employed in the simulator to and will also evaluate left and right turns.

5 Overall Conclusions

The Cooperative Intersection Collision Avoidance System-Stop Sign Assist (CICAS-SSA) is an infrastructure-based driver support system that is intended to improve gap acceptance at rural stop-controlled intersections. The SSA system will track vehicle locations on the major road and then display messages to the driver on the minor road. The primary goal of the current work was to evaluate several candidate CICAS-SSA concepts in order to identify a single sign that may provide the greatest utility in terms of driver performance and usability at a real-world rural intersection. A secondary goal of the current work was to determine the ideal physical characteristics (i.e., location and rotation of a sign relative to drivers) of the candidate CICAS-SSA at a test intersection to maximize comprehension (and subsequent use) of the sign.

The primary goal was accomplished by conducting three studies. The first two studies examined icon use and word selection within several candidate CICAS-SSA signs. The conduct of these studies provided a justification for the redesign of the candidate signs and the elimination of alternative design recommendations. Results of this work indicated:

- Prohibitive messages or messages that provided clear warnings resulted in the highest comprehension rates,
- An Icon sign produced the highest comprehension rates of all signs tested (the icon's "do not cross/turn left" design had a much higher comprehension rate [40%]),
- The Countdown sign's "do not cross/turn left" message resulted in a <10% comprehension rate.
- The Icon and Countdown signs resulted in the highest comprehension of all CICAS-SSA tested.

A third study evaluated driving performance and usability for three candidate SSA sign designs (i.e., icon, countdown, and hazard) compared to a baseline condition for the purpose of identifying the final candidate sign to be field tested at the Minnesota test intersection. Results of this work indicated:

- The presence of CICAS-SSA signs affirms good decision making while also supporting drivers who may have difficulty in the selection and acceptance of a safe gap when crossing.
- For the Countdown and Icon signs, drivers reported using the CICAS-SSA information to help with their crossing decisions.
- Performance data showed that the Countdown sign also resulted in behaviors that may elevate the risk of a crash, such as one-stage crossing maneuvers and misuse of the timer functionality.
- Although a similar distribution of unsafe gaps were rejected while using the Icon sign, participants chose gaps with larger safety margins and exhibited more two-stage crossing maneuvers with this sign compared to the Countdown condition.
- It is recommended that the Icon sign be implemented in an experimental field test.

The secondary goal was to determine the optimal physical characteristics (i.e., location and rotation of a sign relative to drivers) in order to maximize driver comprehension. Results of the work examining sign location and rotation indicated:

- A CICAS-SSA sign that was placed on the shoulder of the near side road on the left side (for the driver positioned at the stop sign) along with a second sign located in the median in front and to the right of the driver (for a driver positioned in the median) was most preferred and resulted in adequate understanding.
- Observations of these sign locations at an actual intersection suggested that visibility of the signs may be poor and the potential of the signs to obscure expressway traffic was highly probably; especially for those drivers seated in larger vehicles (e.g., heavy trucks).
- It was decided that for drivers at the stop sign a CICAS-SSA sign is best positioned in the left-side median.
- It was decided that for drivers in the median a CICAS-SSA sign is best positioned on the far right shoulder.
- A CICAS-SSA sign placed parallel to the mainline roadway was associated with a high degree of comprehension (i.e., drawing a clear association between the sign information and the roadway to which it applied), however, this angle also proved difficult to view.
- A CICAS-SSA sign that was placed parallel to the minor roadway (i.e., directly facing a driver) was easy to view but was also associated with suboptimal comprehension.
- The 45 degree angle did not produce any errors in comprehension, was reported as comfortable and easy to view, and was preferred by over 75% of the respondents.
- It was recommended that a 45 degree angle (or similar) be implemented in further testing.

5.1 Benefits of Validation in a Simulated Driving Context

Using a simulated driving context provides useful CICAS-SSA sign information relative to content, placement, and usage in a cost-effective method. A central consideration in simulation-based studies is the degree to which data collected in the simulation environment will be representative of that observed in real-world settings. No virtual environment can claim to completely reproduce the visual information or proprioceptive cues found in a real-world setting. For example, judgments of distance or vehicle arrival are often compressed as compared to real world estimates (Hancock & Manser, 1997; Manser & Hancock, 1996; Witmer & Sadowski, 1988). Perceived distances have an effect on speed estimation and since the limitations of the motion quality in virtual environments is not perfectly matched with actual vehicles, results from simulator studies should be used to compare relative differences between conditions tested within the same simulator environments and scenarios. Significant findings in simulated environments can be applied to real-world settings by realizing that the direction of any effects seen (i.e. relative validity) are more relevant than actual performance seen in the virtual world (i.e. absolute validity). This has been shown when comparing speed and lane position performance in similar simulated and real tunnel environments (Tornros, 1998). Tornros found that absolute behavioral validity for speed choice was not quite satisfactory, although relative validity was good for speed and lane position between the simulated and real-world environments. Likewise Godley, Triggs and Fildes (2002) measured speed performance during stopping maneuvers and curves in response to real and virtual rumble strips. Their results also support the notion that relative and not absolute behavioral patterns exist between virtual and real-world behavior.

One method for examining relative validity between CICAS-SSA signs in a simulated environment to promote generalization between simulation and actual roadway environments is to include elements from real-world environments in simulation studies. This has been performed with respect to driving scenes, scenarios, and metrics. Relative to simulated driving scenes, the simulation study described here employed 3-D scenes accurate to 1 cm which were representative of the actual test intersection. Within these scenes, objects (i.e., signs, guideposts, lane markings, etc) were positioned in locations that were identical to the actual intersection. Driving scenarios from the actual test intersection were also replicated in the simulated environment. Specifically, traffic flow data at the test intersection were obtained from a previous on-road test (Gorjestani, et al). This data yielded information that delineated the frequency of observed average gap sizes passing the intersection throughout each day. The observed gap thresholds were then employed in simulation studies that required the driver to assess safe and unsafe gap thresholds. When designing the simulation studies presented here care was taken to use metrics that will be analyzed in the field test.

References

AASHTO Strategic Highway Safety Plan. American Association of State Highway and Transportation Officials. Washington, D.C., 1998.

Campbell, J.L., Richman, J.B., Carney, C., & Lee, J.D. (2004a). *In-vehicle display icons and other information elements volume 1: Guidelines* (Rep. No. FHWA-RD-03-065). McLean, VA: Federal Highway Administration.

Campbell, J.L., Hoffmeister, D.H., Keifer, R.J., Selke, D.J., Green, P.A., & Richman, J.B. (2004b)."Comprehension Testing of Active Safety Symbols" (2004-01-0450). *2004 SAE World Congress and Exhibition Technical Papers*, Detroit, MI, March 8-11, 2004.

Chrysler, S.T., Wright, J., & Williams, A. (2004). *3D Visualization as a Tool to Evaluate Sign Comprehension* (Rep. No. SWUTC/04/167721-1). College Station, TX: Southwest University Transportation Center.

Creaser, J.I., Rakauskas, M.E., Ward, N.J., Laberge, J.C., & Donath, M. (2007). "Concept evaluation of intersection decision support (IDS) system interfaces to support drivers' gap acceptance decisions at rural stop-controlled intersections." *Transportation Research Part F, 10*, 208-228.

Dewar, R.E., Kline, D.W., & Swanson, H.A. (1994). "Age differences in comprehension of traffic sign symbols." *Transportation Research Record 1456*, 1-10.

Federal Highway Administration (2004). "Intersection safety facts and statistics." US Department of Transportation: FHWA. http://safety.fhwa.dot.gov/intersections/inter_facts.htm (accessed January 23, 2008).

Godley, S.T., Triggs, T.J., & Fildes, B.N. (2002). "Driving simulator validation for speed research." *Accident Analysis and Prevention*, 34, 589-600.

Gorjestani A., Menon, A., Cheng, P., Shankwitz, C. & Donath, M., (2010) *Determination of the Alert and Warning Timing for the Cooperative Intersection Collision Avoidance System – Stop Sign Assist Using Macroscopic and Microscopic Data: CICAS-SSA Report #1*, August.

Laberge, J.C., Creaser, J.I., Rakauskas, M.E., & Ward, N.J. (2006). "Design of an intersection decision support (IDS) interface to reduce crashes at rural stop-controlled intersections." *Transportation Research Part C, 14*, 39-56.

Neuman, T., Pfefer, R., Slack, K., Harwood, D., Potts, I., Torbic, D., & Rabbani, E. (2003) *NCHRP Report 500 (Volume 5): A Guide for Addressing Unsignalized Intersection Collisions*, TRB, National Research Council, Washington, D.C.

Preston, H., & Storm, R. (2003). "Review of Minnesota's rural crash data." Draft-Technical Memorandum, 112003. Eagan, MN: CH2M HILL, October.

Preston, H., Storm, R., Donath, M., & Shankwitz, C. (2004). *Review of Minnesota's rural crash data: Methodology for identifying intersections for intersection decision support (IDS)*. (Rep No. MN/RC-2004-31) St. Paul, MN: Minnesota Department of Transportation.

Shinar, D., Dewar, R.E., Summala, H., & Zakowska, L. (2003). "Traffic sign symbol comprehension: a cross-cultural study." *Ergonomics, 46*(15), 1549-1565.

Stemler, S.E. (2004). "A comparison of consensus, consistency, and measurement approaches to estimating interrater reliability." *Practical Assessment, Research & Evaluation, 9*, Retrieved May 24, 2007 from http://PAREonline.net/.

Tornros, J. (1998). "Driving behaviour in a real and a simulated road tunnel- a validation study." *Accident Analysis and Prevention*, 30(4), 497-503.

Wierwille, W.W. & Casali, J.G. (1983). „A validated rating scale for global mental workload measurement applications." *Proceedings of the 27ths Annual Meeting of the Human Factors and Ergonomics Society*. 129-133. Santa Monica, CA: Human Factors Society.

Witmer, B., & Sadowski, W.J. Jr. (1988). "Nonvisually guided locomotion to a previously viewed target in real and virtual environments." *Human Factors* 40, 478–488.

APPENDIX A. TEST BOOKLET

PART 1

DEMOGRAPHIC QUESTIONNAIRE

The purpose of this questionnaire is to assess your driving experience and obtain background information. Your personal identity will not be associated with any of your responses. Only a unique number will be recorded and will be used by the researchers.

Please complete each question by responding in the space provided or selecting the appropriate response.

5.1.1.1.1

Part I. Demographic Information

1. Are you? ☐ Male **(1)** ☐ Female **(2)**

2. What is your age? _____ years

3. What is your current employment status? ☐ Full Time **(1)** ☐ Part Time **(2)**

☐ Retired **(3)** ☐ Student **(4)**

☐ Unemployed **(5)** ☐ Other: _____ **(6)**

4. Where do you currently live? ☐ Rural area **(1)** ☐ Urban area **(2)**

☐ Suburban area **(3)** ☐ Other: _____ **(4)**

5. How many years have you had your driver's license (excluding learner's permit)?

_____ year(s)

6. On average, how many miles do you drive per year?_____ miles / year

7. How often did you drive last month?

☐ (1)   ☐ (2)   ☐ (3)   ☐ (4)   ☐ (5)

Never   Rarely   Sometimes   Most Days   Every Day

8. Do you drive frequently on *Highways*?   ☐ Yes(1)   ☐ No(2)

9. Do you drive frequently on *Urban Roads*?   ☐ Yes(1)   ☐ No(2)

10. Do you drive frequently on *Rural Roads*?   ☐ Yes(1)   ☐ No(2)

11. In the last 5 years, have you ever been   ☐ Yes(1)   ☐ No(2)
     the driver in a motor-vehicle accident

     If yes, how many *minor* road accidents have you been involved in?____

     A minor accident is one in which no-one required medical treatment AND costs of damage to vehicles and property were less than $1000

Part II. Driving Experience

If yes, how many *major* road accidents have you been involved in?____

A major accident is one in which EITHER someone required medical treatment OR costs of damage to vehicles and property were greater than $1000, or both

If yes, how many times were you cited as being at fault in the accident?_____

12. What type of vehicle do you drive most often (check one)?

☐ Motorcycle **(1)**   ☐ Passenger Car **(2)**

☐ Pick-Up Truck **(3)**   ☐ Sport utility vehicle **(4)**

☐ Van or Minivan **(5)**   ☐ Other:_____ **(6)**

13. How would you rate your driving skill compared to your peers?

☐(1)	☐(2)	☐(3)	☐(4)	☐(5)
Very bad driver	Bad driver	Average driver	Good driver	Very good driver

14. How would you rate your overall health compared to your peers?

☐(1)	☐(2)	☐(3)	☐(4)	☐(5)
Very Poor	Poor	Average	Good	Excellent

Participant ID:____
Date:____

PART 2

In this part of the study, you will be provided with a driving scenario and a picture of one sign on each page. Please read the driving scenario carefully before answering each question. Each driving scenario is a situation that could occur at the type of intersection described to you by the researcher and shown to you in the video. If you need a reminder of what this type of intersection looks like, you may refer to the Study Introduction (laminated sheet).

Please work through this booklet <u>one page at a time.</u> You may only move onto Part 3 once you have completely finished Part 2. Please do not flip ahead.

Please try your best to provide an answer for each question.

If you have any questions while completing this booklet please ask the researcher.

You may begin when you are ready.

DRIVING SCENARIO

Think about the intersection that was described to you during the study introduction.

Imagine you are in your car and <u>you are stopped at the STOP sign on the minor road waiting to cross the intersection at the major road.</u>

A "smart" sign exists at this intersection. The information displayed on this sign changes in real time depending on the flow of traffic near the intersection. The messages on the sign below correspond to the traffic conditions present at the intersection right now.

For the sign pictured, please describe <u>in your own words</u> what you think this sign means for this **driving scenario** described above:

_____
_____
_____

DRIVING SCENARIO

Think about the intersection that was described to you during the study introduction.

Imagine you are in your car and <u>you are stopped at the STOP sign on the minor road waiting to cross the intersection at the major road.</u>

A "smart" sign exists at this intersection. The information displayed on this sign changes in real time depending on the flow of traffic near the intersection. The messages on the sign below correspond to the traffic conditions present at the intersection right now.

For the sign pictured, please describe <u>in your own words</u> what you think this sign means for this **driving scenario** described above:

___
___
___

DRIVING SCENARIO

Think about the intersection that was described to you during the study introduction.

Imagine you are in your car and <u>you are stopped at the STOP sign on the minor road waiting to cross the intersection at the major road.</u>

A "smart" sign exists at this intersection. The information displayed on the <u>yellow</u> portion of this sign changes in real time depending on the flow of traffic near the intersection. The messages on the sign below correspond to the traffic conditions present at the intersection right now.

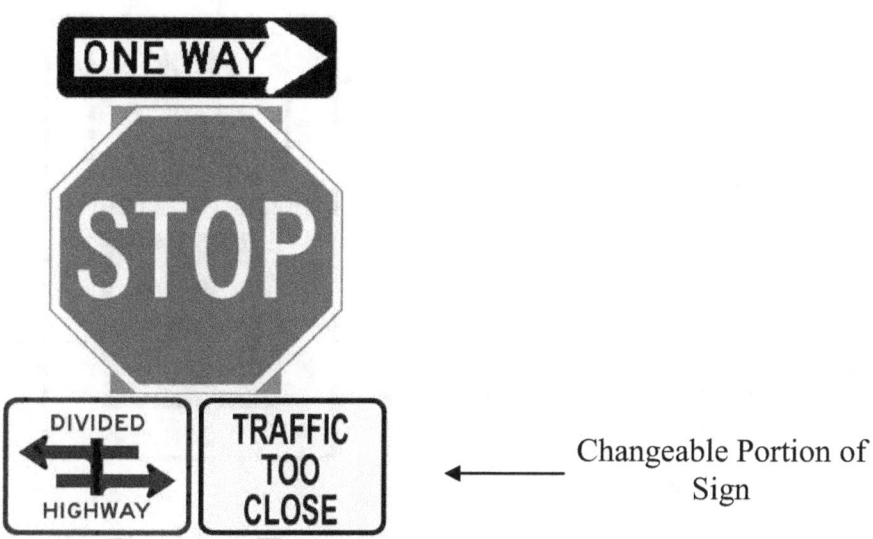

← Changeable Portion of Sign

For the sign pictured, please describe <u>in your own words</u> what you think this sign means for this **driving scenario** described above:

_____
_____
_____

DRIVING SCENARIO

Think about the intersection that was described to you during the study introduction.

Imagine you are in your car and you are stopped at the STOP sign on the minor road waiting to cross the intersection at the major road.

A "smart" sign exists at this intersection. The information displayed on this sign changes in real time depending on the flow of traffic near the intersection. The messages on the sign below correspond to the traffic conditions present at the intersection right now.

For the sign pictured, please describe in your own words what you think this sign means for this **driving scenario** described above:

_____
_____
_____

DRIVING SCENARIO

Think about the intersection that was described to you during the study introduction.

Imagine you are in your car and <u>you are stopped at the STOP sign on the minor road waiting to cross the intersection at the major road.</u>

A "smart" sign exists at this intersection. The information displayed on this sign changes in real time depending on the flow of traffic near the intersection. The messages on the sign below correspond to the traffic conditions present at the intersection right now.

For the sign pictured, please describe <u>in your own words</u> what you think this sign means for this **driving scenario** described above:

_____
_____
_____

DRIVING SCENARIO

Think about the intersection that was described to you during the study introduction.

Imagine you are in your car and <u>you are stopped at the STOP sign on the minor road waiting to cross the intersection at the major road.</u>

A "smart" sign exists at this intersection. The information displayed on the <u>yellow</u> portion of this sign changes in real time depending on the flow of traffic near the intersection. The messages on the sign below correspond to the traffic conditions present at the intersection right now.

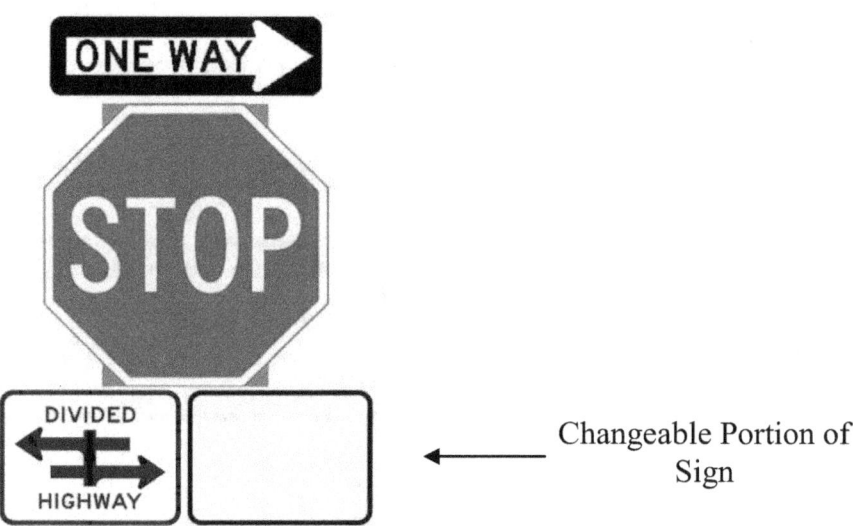

← Changeable Portion of Sign

For the sign pictured, please describe <u>in your own words</u> what you think this sign means for this **driving scenario** described above:

_____
_____
_____

A-10

DRIVING SCENARIO

Think about the intersection that was described to you during the study introduction.

Imagine you are in your car and <u>you are stopped at the STOP sign on the minor road waiting to cross the intersection at the major road.</u>

A "smart" sign exists at this intersection. The information displayed on this sign changes in real time depending on the flow of traffic near the intersection. The messages on the sign below correspond to the traffic conditions present at the intersection right now.

For the sign pictured, please describe <u>in your own words</u> what you think this sign means for this **driving scenario** described above:

_____
_____
_____

DRIVING SCENARIO

Think about the intersection that was described to you during the study introduction.

Imagine you are in your car and <u>you are stopped at the STOP sign on the minor road waiting to cross the intersection at the major road.</u>

A "smart" sign exists at this intersection. The information displayed on this sign changes in real time depending on the flow of traffic near the intersection. The messages on the sign below correspond to the traffic conditions present at the intersection right now.

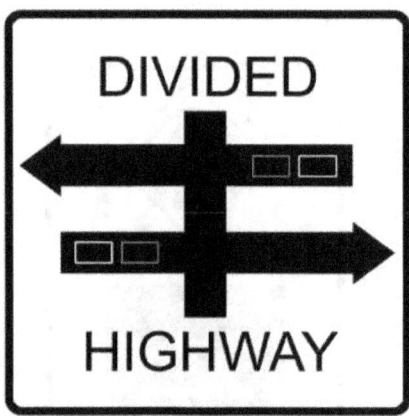

For the sign pictured, please describe <u>in your own words</u> what you think this sign means for this **driving scenario** described above:

_____
_____
_____

DRIVING SCENARIO

Think about the intersection that was described to you during the study introduction.

Imagine you are in your car and <u>you are stopped at the STOP sign on the minor road waiting to cross the intersection at the major road.</u>

A "smart" sign exists at this intersection. The information displayed on this sign changes in real time depending on the flow of traffic near the intersection. The messages on the sign below correspond to the traffic conditions present at the intersection right now.

For the sign pictured, please describe <u>in your own words</u> what you think this sign means for this **driving scenario** described above:

_____
_____
_____

DRIVING SCENARIO

Think about the intersection that was described to you during the study introduction.

Imagine you are in your car and you are stopped at the STOP sign on the minor road waiting to cross the intersection at the major road.

A "smart" sign exists at this intersection. The information displayed on this sign changes in real time depending on the flow of traffic near the intersection. The messages on the sign below correspond to the traffic conditions present at the intersection right now.

For the sign pictured, please describe in your own words what you think this sign means for this **driving scenario** described above:

_____
_____
_____

DRIVING SCENARIO

Think about the intersection that was described to you during the study introduction.

Imagine you are in your car and you are stopped at the STOP sign on the minor road waiting to cross the intersection at the major road.

A "smart" sign exists at this intersection. The information displayed on this sign changes in real time depending on the flow of traffic near the intersection. The messages on the sign below correspond to the traffic conditions present at the intersection right now.

For the sign pictured, please describe in your own words what you think this sign means for this **driving scenario** described above:

_____
_____
_____

DRIVING SCENARIO

Think about the intersection that was described to you during the study introduction.

Imagine you are in your car and you are stopped at the STOP sign on the minor road waiting to cross the intersection at the major road.

A "smart" sign exists at this intersection. The information displayed on this sign changes in real time depending on the flow of traffic near the intersection. The messages on the sign below correspond to the traffic conditions present at the intersection right now.

For the sign pictured, please describe in your own words what you think this sign means for this **driving scenario** described above:

_____
_____
_____

DRIVING SCENARIO

Think about the intersection that was described to you during the study introduction.

Imagine you are in your car and <u>you are stopped at the STOP sign on the minor road waiting to cross the intersection at the major road.</u>

A "smart" sign exists at this intersection. The information displayed on this sign changes in real time depending on the flow of traffic near the intersection. The messages on the sign below correspond to the traffic conditions present at the intersection right now.

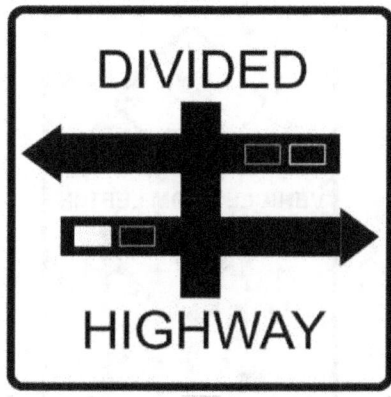

For the sign pictured, please describe <u>in your own words</u> what you think this sign means for this **driving scenario** described above:

_____
_____
_____

DRIVING SCENARIO

Think about the intersection that was described to you during the study introduction.

Imagine you are in your car and you are stopped at the STOP sign on the minor road waiting to cross the intersection at the major road.

A "smart" sign exists at this intersection. The information displayed on this sign changes in real time depending on the flow of traffic near the intersection. The messages on the sign below correspond to the traffic conditions present at the intersection right now.

For the sign pictured, please describe in your own words what you think this sign means for this **driving scenario** described above:

_____
_____
_____

Participant ID:____
Date:____

PART 3

In this part of the study, you will be provided with a driving scenario and a picture of more than one sign. The driving scenario in this part will also include a description of what the signs shown are intended to mean. Please carefully read each driving scenario and sign description before answering the question.

Please work through this part one page at a time.

If you have any questions while completing this part of the booklet please ask the researcher.

You may not return to Part 2 once you have begun Part 3.

You may begin when you are ready.

DRIVING SCENARIO

Think about the intersection that was described to you during the study introduction.

Imagine you are in your car and <u>you are stopped at the STOP sign on the minor road waiting to cross the intersection at the major road.</u>

A "smart" sign exists at this intersection. The information displayed on this sign changes in real time depending on the flow of traffic near the intersection. The messages on the sign below correspond to the traffic conditions present at the intersection right now.

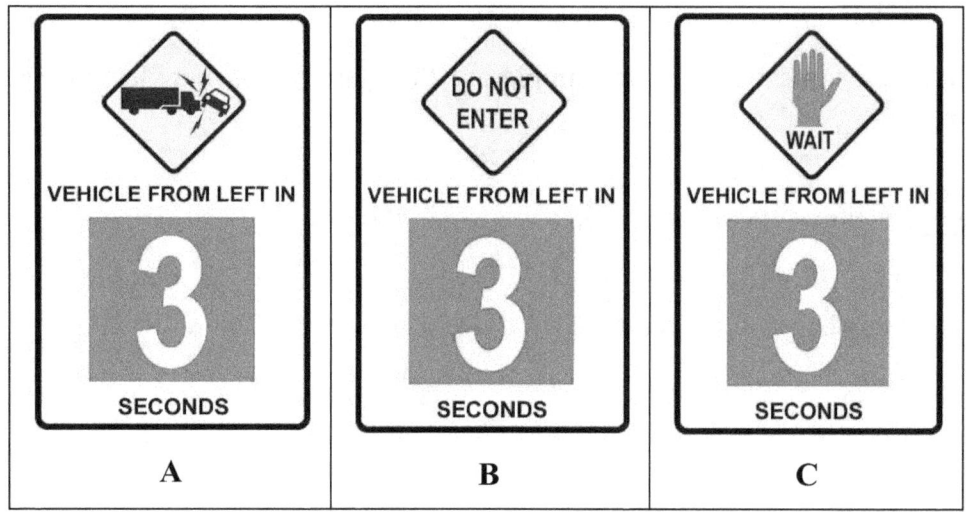

The above signs are intended to provide the following message:

"It is not safe to enter the intersection; traffic approaching from both directions. A vehicle is approaching from the left and is 3 s away."

Please **rank order** each version (A, B, C) of this sign based on <u>how accurately</u> you think each version conveys the above message related to the driving scenario. A rank of "1" means the version <u>most accurately</u> conveys the message described. A rank of "3" means the version <u>least accurately</u> conveys the message described. Put the letter of each version next to the corresponding rank you chose for the version.

Rankings:

1_____

A-20

2 _____
3 _____

Please explain your choices:

_____
_____

_____

DRIVING SCENARIO

Think about the intersection that was described to you during the study introduction.

Imagine you are in your car and <u>you are stopped at the STOP sign on the minor road waiting to cross the intersection at the major road.</u>

A "smart" sign exists at this intersection. The information displayed on this sign changes in real time depending on the flow of traffic near the intersection. The messages on the sign below correspond to the traffic conditions present at the intersection right now.

The signs above are intended to provide the following message:

"There is traffic in the far lanes and it may not be safe to cross or turn left into the far lanes. However, it might be safe to turn right because the vehicle from left is 11 s away."

Please **rank order** each version (A, B) of this sign based on how accurately you think each version conveys the above message related to the driving scenario. A rank of "1" means the version most accurately conveys the message described. A rank of "2" means the version least accurately conveys the message described. Put the letter of each version next to the corresponding rank you chose for the version.

Rankings:
1_____
2_____

Please explain your choices:

_____

_____

_____

DRIVING SCENARIO

Think about the intersection that was described to you during the study introduction.

Imagine you are in your car and you are stopped at the STOP sign on the minor road waiting to cross the intersection at the major road.

A "smart" sign exists at this intersection. The information displayed on this sign changes in real time depending on the flow of traffic near the intersection. The messages on the sign below correspond to the traffic conditions present at the intersection right now.

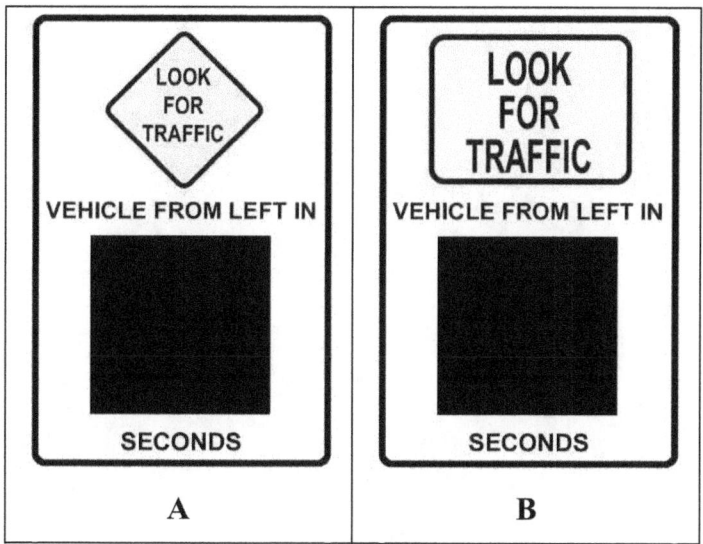

The signs above are intended to provide the following message:

"It may be safe to cross or make a turn into the intersection, but you should still watch for approaching traffic."

Please **rank order** each version (A, B) of this sign based on how accurately you think each version conveys the above message related to the driving scenario. A rank of "1" means the version most accurately conveys the message described. A rank of "2" means the version least accurately conveys the message described. Put the letter of each version next to the corresponding rank you chose for the version.

Rankings:

1_____
2_____

Please explain your choices:

APPENDIX B. CONSENT FORM

CONSENT FORM

Sign Comprehension Study: Experiment One

You are invited to be in a research study to examine the understandability of new sign designs for use in an intersection decision support system. You were selected a possible participant because you responded to our ads requesting participants and were found to be a suitable participant for this study. We ask that you read this form carefully and ask any questions you may have before agreeing to be in the study.

This study is being conducted by Janet Creaser and Nic Ward who are research scientists in the HumanFIRST Program at the University of Minnesota.

Background Information:

The purpose of this study is to investigate how well drivers comprehend new sign designs that will be used in an intersection decision support system.

Procedures:

If you agree to be in this study, we will ask you to do the following things: (1) provide us with some basic information about yourself and your driving history (e.g., age, number of years you have had your license); (2) view a video of a vehicle driving through an intersection that is typical of where the types of signs you will view today will be located; (3) view a number of sign configurations and write in your own words what you think each sign means after you have examined it; (4) rank order different designs based on how well they match the description of what they are intended to tell you. The total time to complete this study today is about 1 hour.

Risks and Benefits of Being in the Study:

There are no direct benefits to you for participating in this study. There are no risks associated with participating in this study.

Compensation:

You will receive a payment of $40 for participation. If you terminate the study early, you will still receive full payment.

Confidentiality:

The records of this study will be kept private. You name will not be associated with any of the data collected today. In any sort of report we might publish, we will not include any information that will make it possible to identify you or other participants. Research records are stored securely in locked offices and only researchers on this study will have access to the data collected.

Research Results:

The results of this research will be published at the end of the study. If you are interested in obtaining this information, please visit our website (listed at the bottom of this form) for more information.

Voluntary Nature of the Study:

Participation in this study is voluntary. Your decision whether or not to participate will not affect your current or future relations with the University of Minnesota. If you decide to participate, you are free to not answer any question or withdraw at any time without affecting those relationships.

Contacts and Questions:

The researchers conducting this study are Janet Creaser and Nic Ward. You may ask any questions you have now. If you have questions later, **you are encouraged** to contact them at 1100 Mechanical Engineering, 111 Church St SE, Minneapolis, MN, 55455; 612-624-2877; janetc@me.umn.edu.

If you have any questions or concerns regarding this study and would like to talk to someone other than the researcher(s), **you are encouraged** to contact the University of Minnesota's Research Subjects' Advocate Line, D528 Mayo, 420 Delaware St. Southeast, Minneapolis, Minnesota 55455; (612) 625-1650.

You will be given a copy of this information to keep for your records.

Statement of Consent:

I have read the above information. I have asked questions and have received answers. I consent to participate in the study.

Signature:_____ Date:_____

Signature of Investigator:_____ Date:_____

APPENDIX C. STUDY INSTRUCTIONS

Note: Read by Researcher to participants after informed consent

"Now that we have read the study introduction, I will explain the booklet to you. The booklet in front of you contains all the information you need to complete the study. The booklet contains three parts.

Part 1

"The first part of the booklet is the Driver Questionnaire. This simply provides us with background information about your experience with driving."

Part 2

"On each page of the booklet in part 2, you will be provided with a driving scenario and shown a sign. Your task is to write in your own words what you think this sign means based on the driving scenario that has been presented to you. It is important that you read each driving scenario carefully. Please try to answer the questions to the best of your ability.

The signs that you will see today in this booklet are "smart" signs and would be located near the stop sign at the intersection. The messages on these signs can change depending on the flow traffic on the cross road. These messages provide information to the driver about the traffic approaching the intersection."

Part 3

"In Part 3, you will be provided with a driving scenario and a description of how a sign is intended to work. Please read the sign description carefully. You will use this description to rank order signs based on how well you think each one conveys the information in the description."

"Written instructions are also be provided in the booklet to help you complete each part. If you have any questions while you are completing the booklet please ask."

APPENDIX D. SCORING CRITERIA FOR SIGNS

Sign		Rating Criteria
S1	(sign image: VEHICLE FROM LEFT IN 3 SECONDS)	**Meaning:** Do not enter because traffic is too close **Major Information Elements:** - Time (lag) - Unsafe threshold (red means car within unsafe time) - Crash icon - Diamond means hazard/caution **Minor Information Elements:** - Direction of approaching traffic (left)
S2	(sign image: DO NOT ENTER VEHICLE FROM LEFT IN 5 SECONDS)	**Meaning:** Do not enter because traffic is too close **Major Information Elements:** - Time (lag) - Unsafe threshold (red means car within unsafe time) - Diamond means hazard/caution **Minor Information Elements:** - Direction of approaching traffic (left)
S3	(sign image: VEHICLE FROM LEFT IN 4 SECONDS)	**Meaning:** Do not enter because traffic is too close **Major Information Elements:** - time (lag) - Unsafe threshold (red means car within unsafe time) - Wait hand and text - Diamond means hazard/caution **Minor Information Elements:** - Direction of approaching traffic (left)
S4	(sign image: VEHICLE FROM LEFT IN SECONDS)	**Meaning:** May be ok to turn right because no vehicles detected in near lanes, but do not cross or turn left beyond median (straight/left) because far lanes traffic is close **Major Information Elements:** - No time information means no car approaching/detected in near lanes - Threshold (black means vehicle not detected inside unsafe threshold) - Prohibited to go straight/turn left - Diamond means hazard/caution **Minor Information Elements:** - Direction of approaching traffic (left)

S5	[sign: arrow straight/right, "VEHICLE FROM LEFT IN 11 SECONDS"]	**Meaning:** May be ok to turn right because no vehicles detected in near lanes, but do not cross or turn left beyond median (straight/left) because far lanes traffic is close **Major Information Elements:** • No time information means no car approaching/detected in near lanes • Threshold (black means vehicle not detected inside unsafe threshold) • Prohibited to go straight/turn left **Minor Information Elements:** • Direction of approaching traffic (left)
S6	[sign: "LOOK FOR TRAFFIC, VEHICLE FROM LEFT IN 10 SECONDS"]	**Meaning:** Proceed with caution because vehicle from left is detected but far enough away **Major Information Elements:** • Time (lag) • Threshold (black means vehicle detected outside unsafe threshold) • Look for traffic text • Yellow color means caution **Minor Information Elements:** • Direction of approaching traffic (left)
S7	[diamond sign: "LOOK FOR TRAFFIC, VEHICLE FROM LEFT IN _ SECONDS"]	**Meaning:** Proceed with caution; no vehicles detected near intersection **Major Information Elements:** • Time (lag) • Threshold (black means vehicle not detected inside unsafe threshold) • Look for traffic text • Diamond means caution **Minor Information Elements:** • Direction of approaching traffic (left)
S8	[sign: "DIVIDED HIGHWAY" with arrows]	**Meaning:** Do not enter (cross/turn) because traffic is too close in near lanes **Major Information Elements:** • Car approaching from left too close (red marker filled) • Prohibited to go straight/right for near lanes **Minor Information Elements:** • Near and far lanes yellow car markers (unfilled) • Far lanes red car marker (unfilled)

			• Direction of lanes • Type of highway
S9		[DIVIDED HIGHWAY sign]	**Meaning:** Do not enter (cross/turn) because traffic is too close in both near and far lanes **Major Information Elements:** • Car approaching from right too close (red marker filled) • Car approaching from left too close (red marker filled) • Prohibited to go straight/left/right for both sets of lanes **Minor Information Elements:** • Near and far lanes yellow car markers (unfilled) • Direction of lanes • Type of highway
S10		[DIVIDED HIGHWAY sign]	**Meaning:** May be ok to turn right because no vehicles detected in near lanes, but do not cross or turn left beyond median (straight/left) because far lanes traffic is close **Major Information Elements:** • Car approaching from right too close (red marker) • Prohibited to go straight/left • Near car red marker (unfilled) **Minor Information Elements:** • Near and far lanes yellow car markers (unfilled) • Direction of lanes • Type of highway
S11		[DIVIDED HIGHWAY sign]	**Meaning:** Proceed with caution (car detected outside unsafe threshold from left) **Major Information Elements:** • Car approaching from left (yellow marker filled) • Red car markers for near and far lanes (unfilled) **Minor Information Elements:** • Far lanes yellow car marker (unfilled) • Direction of lanes • Type of highway
S12		[DIVIDED HIGHWAY sign]	**Meaning:** No traffic detected in near/far lanes; proceed into intersection **Major Information Elements:** • Near car red markers (unfilled) **Minor Information Elements:**

		• Far car yellow markers (unfilled) • Direction of lanes • Type of highway
S13	[ONE WAY STOP / DIVIDED HIGHWAY / TRAFFIC TOO CLOSE sign image]	**Meaning:** Traffic detected too close to intersection; do not proceed; **Major Information Elements:** • Yellow caution color • Traffic too close text **Minor Information Elements:** • None
S14	[ONE WAY STOP / DIVIDED HIGHWAY sign image]	**Meaning:** No traffic detected within unsafe threshold; proceed with caution **Major Information Elements:** • Yellow caution color **Minor Information Elements:** • None

APPENDIX E. KAPPA RESULTS FOR INTER-RATER RELIABILITY

Cohen's kappa was also run on the scores for each sign to ensure that, statistically, the level of agreement between researchers was appropriate. The results of the analysis showed that all levels of agreement were statistically significant at p<0.001 (see Table). However, a rule of thumb with Cohen's kappa is that values of 0.41-0.60 indicate moderate levels of agreement among raters and that values above 0.60 are substantial (Stemler, 2004). In our analyses, interrater reliability using Cohen's kappa was greater than 0.60 for two signs and was between 0.41-0.60 for six signs. The remaining six signs all had values above 0.30 that were statistically significant. The ratings for the signs with kappa values below 0.4 were reviewed to determine if the differences were due to large discrepancies in the application of scoring (e.g., 2 versus 5), or if the discrepancies occurred within the criteria used to determine percent agreement (e.g., 1 vs. 2).

Table E-1. Cohen's kappa for interrater reliability between the researchers.

Sign Type	Sign	Kappa Value	Sign	Kappa Value	Sign	Kappa Value	Sign	Kappa Value
Countdown Signs		0.565*		0.454*		0.542*		0.512*
		0.307* (sign 9)				0.463*		0.33*
Icon Signs`		0.376*		0.314*		0.395*		
		0.331*		0.612*				
Hazard Signs		0.54*		0.708*				

* p<0.001

These six signs showed an 80.6% rate of agreement between the researchers, which is only slightly lower than the overall percent agreement and still within reasonable tolerances for agreement. In comparison, the percent agreement for the eight signs with kappa values greater than 0.40 was 84.6%. An inspection of the paired ratings for the six signs with kappa values below 0.40 showed that most scores differed within the same comprehension level. Therefore, the lower kappa values are mostly attributed to differences in scoring that occurred within a comprehension level rather than from discrepancies across comprehension levels. Overall, the results indicate good interrater reliability for this analysis.

APPENDIX F. ROTATION STUDY, POST-TASK QUESTIONS

Please answer the following questions based on your last experience viewing the sign.

1. With the laser pointer, please indicate to the experimenter which traffic or vehicles the sign is giving information about.

    How confident are you in your answer to the above question?

☐	☐	☐	☐	☐
Not at all confident	Somewhat not confident	Neutral	Somewhat confident	Completely confident

Please mark your agreement with the following statements. The sign I viewed ...

2. Made it easy for me to associate information on the sign to traffic conditions.

☐	☐	☐	☐	☐
Strongly Disagree	Disagree	Neutral	Agree	Strongly Agree

    Please explain: _____
    _____

3. Was comfortable to view in this location.

☐	☐	☐	☐	☐
Strongly Disagree	Disagree	Neutral	Agree	Strongly Agree

    Please explain: _____
    _____

4. Obstructed my view of the approaching traffic on the main road.

☐	☐	☐	☐	☐
Strongly Disagree	Disagree	Neutral	Agree	Strongly Agree

    Please explain: _____

_____

5. Was easy to see at this distance.

☐	☐	☐	☐	☐
Strongly Disagree	Disagree	Neutral	Agree	Strongly Agree

Please explain: _____
_____

APPENDIX G. ROTATION STUDY, POST-VIEWING LOCATION QUESTIONS

Pn Viewing Location, ICON Sign

Please answer the following questions based on your experience viewing all three sign viewing angles at this location.

1. Please <u>rank the images of sign viewing angles</u> from the one that maps to the roadway, from best (1) to worst (3):

Rank: _____   Rank: _____   Rank: _____

2. Why did you prefer the viewing angle you ranked as a "1"?
Please explain in as much detail as possible:

3. Why did you <u>not</u> prefer the viewing angles you ranked as a "2" and "3"?
Please explain in as much detail as possible:

Pn Viewing Location, COUNTDOWN Sign

Please answer the following questions based on your experience viewing all three sign viewing angles at this location.

1. Please <u>rank the images of sign viewing angles</u> from the one that maps to the roadway, from best (1) to worst (3):

Rank: _____   Rank: _____   Rank: _____

2. Why did you prefer the viewing angle you ranked as a "1"?
Please explain in as much detail as possible:

3. Why did you <u>not</u> prefer the viewing angles you ranked as a "2" and "3"?
Please explain in as much detail as possible:

Pf Viewing Location, ICON Sign

Please answer the following questions based on your experience viewing all three sign viewing angles at this location.

1. Please <u>rank the images of sign viewing angles</u> from the one that maps to the roadway, from best (1) to worst (3):

Rank: _____   Rank: _____   Rank: _____

2. Why did you prefer the viewing angle you ranked as a "1"?
Please explain in as much detail as possible:

3. Why did you <u>not</u> prefer the viewing angles you ranked as a "2" and "3"?
Please explain in as much detail as possible:

Pf Viewing Location, COUNTDOWN Sign

Please answer the following questions based on your experience viewing all three sign viewing angles at this location.

1. Please <u>rank the images of sign viewing angles</u> from the one that maps to the roadway, from best (1) to worst (3):

Rank: _____   Rank: _____   Rank: _____

2. Why did you prefer the viewing angle you ranked as a "1"?
Please explain in as much detail as possible:

3. Why did you <u>not</u> prefer the viewing angles you ranked as a "2" and "3"?
Please explain in as much detail as possible:

APPENDIX H. LOCATION STUDY, POST-TASK QUESTIONNAIRES

Pn Viewing Location, Sign Location Set A, Trial 1

O This circle indicates where the signs were located during this condition.

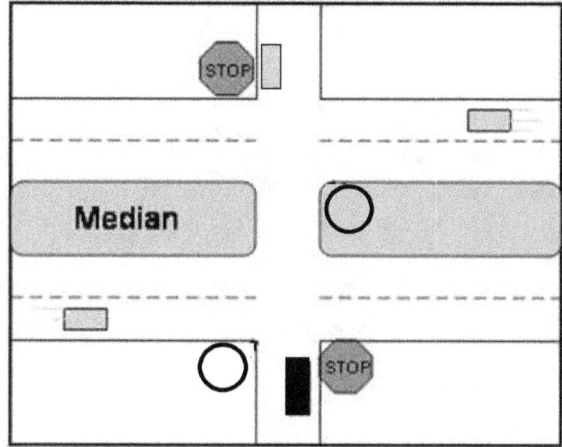

1. At this location, how confident are you that you identified the correct traffic or vehicles the sign was giving information about?

   ☐   ☐   ☐   ☐   ☐

   Not at all confident   Somewhat not confident   Neutral   Somewhat confident   Completely confident

   Please explain: _____

   _____

   _____

Pf Viewing Location, Sign Location Set A, Trial 1

O This circle indicates where the signs were located during this condition.

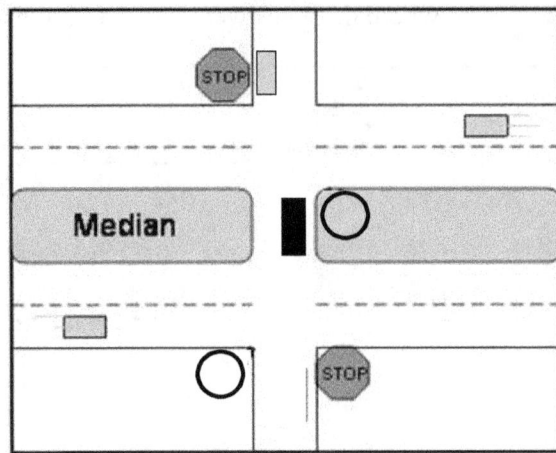

2. At this location, how confident are you that you identified the correct traffic or vehicles the sign was giving information about?

☐	☐	☐	☐	☐
Not at all Confident	Somewhat not confident	Neutral	Somewhat confident	Completely confident

Please explain: _____

_____

_____

H-2

Pn Viewing Location, Sign Location Set A, Trial 2

O This circle indicates where the signs were located during this condition.

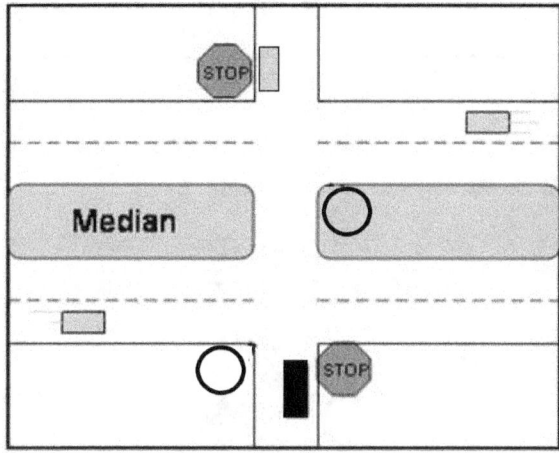

Please indicate your level of agreement with the following statements. **The sign I viewed while at the STOP SIGN...**

3. Made it easy for me to associate information on the sign to traffic conditions.

   ☐ Strongly Disagree   ☐ Disagree   ☐ Neutral   ☐ Agree   ☐ Strongly Agree

   Please explain: _____

4. Was comfortable to view in this location.

   ☐ Strongly Disagree   ☐ Disagree   ☐ Neutral   ☐ Agree   ☐ Strongly Agree

   Please explain: _____

5. Obstructed my view of the approaching traffic on the main road.

   ☐ Strongly Disagree   ☐ Disagree   ☐ Neutral   ☐ Agree   ☐ Strongly Agree

   Please explain: _____

6. Was easy to see at this distance.

   ☐ Strongly Disagree   ☐ Disagree   ☐ Neutral   ☐ Agree   ☐ Strongly Agree

   Please explain: _____

Pf Viewing Location, Sign Location Set A, Trial 2

O This circle indicates where the signs were located during this condition.

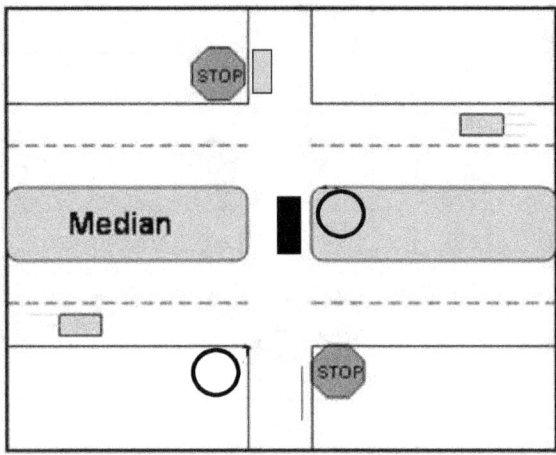

Please indicate your level of agreement with the following statements. **The sign I viewed while in the MEDIAN...**

7. Made it easy for me to associate information on the sign to traffic conditions.

   ☐ Strongly Disagree ☐ Disagree ☐ Neutral ☐ Agree ☐ Strongly Agree

   Please explain: _____

8. Was comfortable to view in this location.

   ☐ Strongly Disagree ☐ Disagree ☐ Neutral ☐ Agree ☐ Strongly Agree

   Please explain: _____

9. Obstructed my view of the approaching traffic on the main road.

   ☐ Strongly Disagree ☐ Disagree ☐ Neutral ☐ Agree ☐ Strongly Agree

   Please explain: _____

10. Was easy to see at this distance.

☐	☐	☐	☐	☐
Strongly Disagree	Disagree	Neutral	Agree	Strongly Agree

Please explain: _____

Pn Viewing Location, Sign Location Set B, Trial 1

O This circle indicates where the signs were located during this condition.

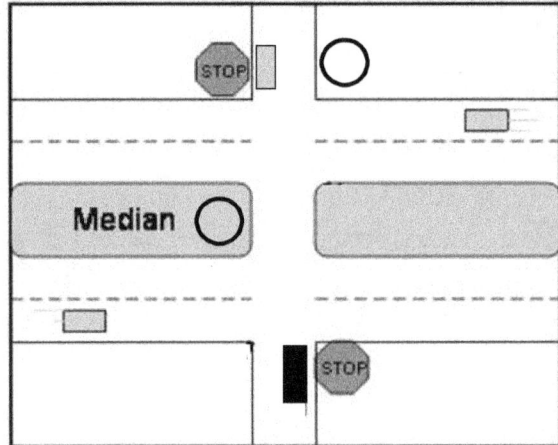

1. At this location, how confident are you that you identified the correct traffic or vehicles the sign was giving information about?

   ☐ Not at all Confident   ☐ Somewhat not confident   ☐ Neutral   ☐ Somewhat confident   ☐ Completely confident

   Please explain: _____

   _____

   _____

Pf Viewing Location, Sign Location Set B, Trial 1

O This circle indicates where the signs were located during this condition.

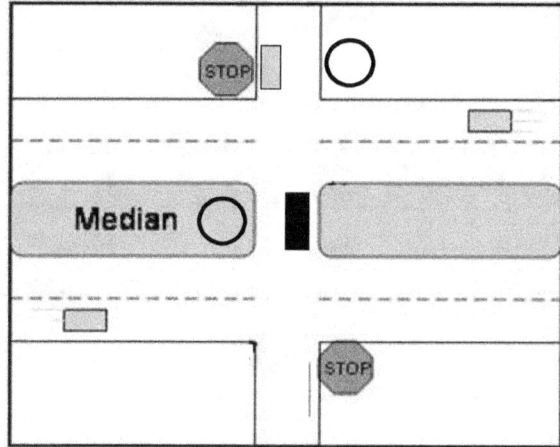

2. At this location, how confident are you that you identified the correct traffic or vehicles the sign was giving information about?

☐	☐	☐	☐	☐
Not at all confident	Somewhat not Confident	Neutral	Somewhat confident	Completely confident

Please explain: _____

_____

_____

H-7

Pn Viewing Location, Sign Location Set B, Trial 2

O This circle indicates where the signs were located during this condition.

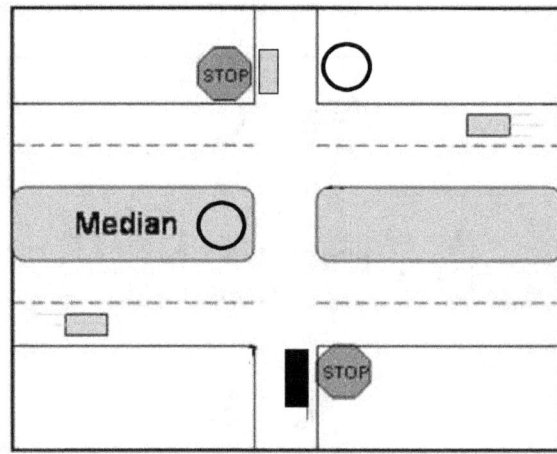

Please indicate your level of agreement with the following statements. **The sign I viewed while at the STOP SIGN…**

3. Made it easy for me to associate information on the sign to traffic conditions.

   ☐ Strongly Disagree   ☐ Disagree   ☐ Neutral   ☐ Agree   ☐ Strongly Agree

   Please explain: _____

4. Was comfortable to view in this location.

   ☐ Strongly Disagree   ☐ Disagree   ☐ Neutral   ☐ Agree   ☐ Strongly Agree

   Please explain: _____

5. Obstructed my view of the approaching traffic on the main road.

   ☐ Strongly Disagree   ☐ Disagree   ☐ Neutral   ☐ Agree   ☐ Strongly Agree

   Please explain: _____

6. Was easy to see at this distance.

   ☐ Strongly Disagree   ☐ Disagree   ☐ Neutral   ☐ Agree   ☐ Strongly Agree

   Please explain: _____

Pf Viewing Location, Sign Location Set B, Trial 2

O This circle indicates where the signs were located during this condition.

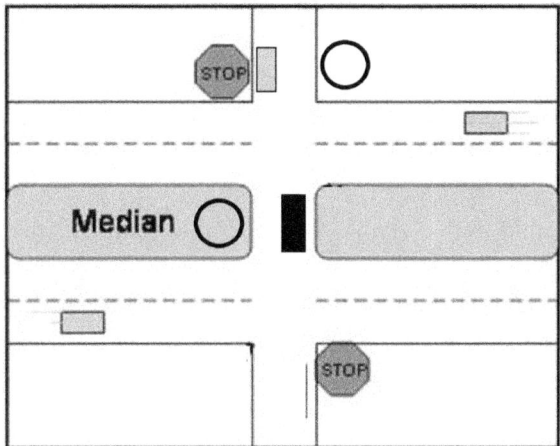

Please indicate your level of agreement with the following statements. **The sign I viewed while in the MEDIAN...**

7. Made it easy for me to associate information on the sign to traffic conditions.
   ☐ ☐ ☐ ☐ ☐
   Strongly Disagree Disagree Neutral Agree Strongly Agree

   Please explain: _____

8. Was comfortable to view in this location.
   ☐ ☐ ☐ ☐ ☐
   Strongly Disagree Disagree Neutral Agree Strongly Agree

   Please explain: _____

9. Obstructed my view of the approaching traffic on the main road.
   ☐ ☐ ☐ ☐ ☐
   Strongly Disagree Disagree Neutral Agree Strongly Agree

Please explain: _____

10. Was easy to see at this distance.

☐	☐	☐	☐	☐
Strongly Disagree	Disagree	Neutral	Agree	Strongly Agree

Please explain: _____

APPENDIX I. LOCATION STUDY, POST-TEST QUESTIONNAIRE

During this study, the signs you viewed were positioned at different locations. The figure below shows the pairs of locations in which the signs were viewed in each drive.

O This circle indicates where the signs were located in layouts A & B.

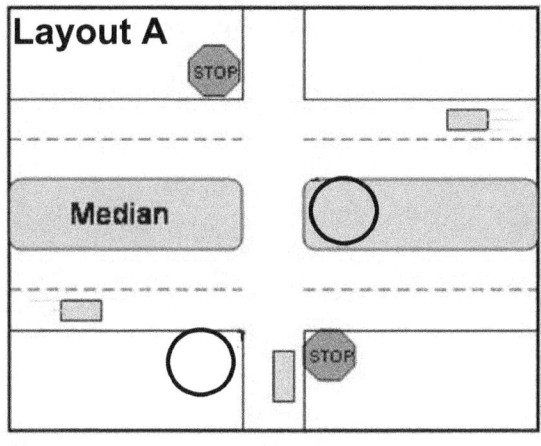
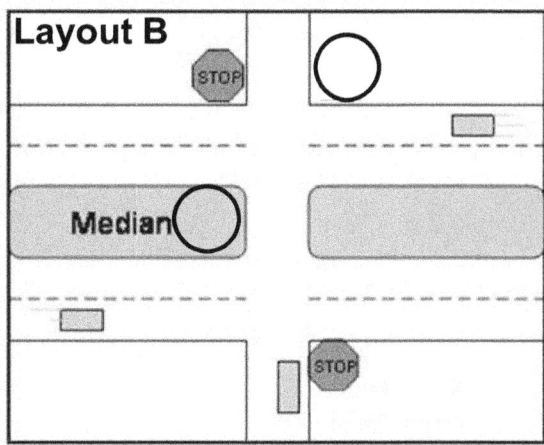

<u>In layout A</u>

The sign you viewed while you were at the stop sign was located immediately to your left.

The sign you viewed while you were in the median was located immediately to your right.

<u>In layout B</u>

The sign you viewed while you were at the stop sign was located across the near lanes in the median.

The sign you viewed while you were in the median was located across the far lanes on the opposite side of the intersection.

1. Based on your experiences of using the signs in each of the locations, which pair of locations did you prefer? Layout …

    ☐ A   ☐ B

2. Why did you prefer this pair of locations? Please explain in as much detail as possible.

The signs you viewed while you were **at the STOP SIGN** were located at one of two positions, as indicated by the O's in the diagram below.

3. Please indicate whether you preferred when the sign was placed ...

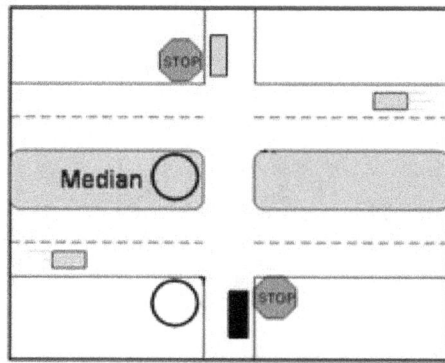

☐ Across the near lanes in the median

☐ Immediately to your left

4. Why did you prefer this location? Please explain in as much detail as possible.

The signs you viewed while you were **in the MEDIAN** were located at one of two positions, as indicated by the O's in the diagram below.

5. Please indicate whether you preferred when the sign was placed ...

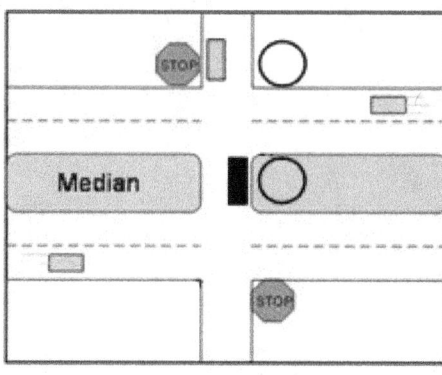

☐ Across the far lanes on the opposite side of the intersection

☐ Immediately to your right

6. Why did you prefer this location? Please explain in as much detail as possible.

APPENDIX J. CONSENT FORM

CONSENT FORM

CICAS Simulator Study Consent Form

You are invited to be in a research study to examine the understandability of new sign designs for use in an intersection decision support system. You were selected a possible participant because you responded to our ads requesting participants and were found to be a suitable participant for this study. We ask that you read this form carefully and ask any questions you may have before agreeing to be in the study.

This study is being conducted by Nicholas Ward, Michael Manser, Janet Creaser and Michael Rakauskas who are research staff in the HumanFIRST Program at the University of Minnesota.

Background Information:

The purpose of this study is to investigate how people drive at intersections in rural environments and how new signs may improve safety at these intersections.

Procedures:

If you agree to be in this study, we will ask you to do the following things: (1) provide permission for us to review your driving record (voluntary); (2) be trained in our driving simulator; and (3) perform a number of directed drives through an intersection in a simulated rural environment of US Highway 52 (US 52). While driving, a head-free eye-gaze tracking system will be used, which may require you to wear reference stickers on your face. You will also be given some questionnaires to complete that ask you about your driving experience and opinion of the signage at the intersection. The duration of the entire study will be about 3 hours.

Risks and Benefits of Being in the Study:

There are no direct benefits to you for participating in this study. A small percentage of individuals may experience motion sickness while driving in the simulator. If you begin to experience this, notify us and we will stop the study. Note: you are free to withdraw from the study at any time if you do not wish to continue.

Compensation:

You will receive a payment of $50 for participation. If you terminate the study early, you will still receive full payment.

Confidentiality:

The records of this study will be kept private. You name will not be associated with any of the data collected today. In any sort of report we might publish, we will not include any information that will make it possible to identify you or other participants. Research records are stored securely in locked offices and only researchers on this study will have access to the data collected.

Research Results:

The results of this research will be published at the end of the study. If you are interested in obtaining this information, please visit our website (listed at the bottom of this form) for more information.

Voluntary Nature of the Study:

Participation in this study is voluntary. Your decision whether or not to participate will not affect your current or future relations with the University of Minnesota. If you decide to participate, you are free to not answer any question or withdraw at any time without affecting those relationships.

Contacts and Questions:

You may ask any questions you have now. If you have questions later, **you are encouraged** to contact Michael Manser by mail at 1101 Mechanical Engineering, 111 Church St SE, Minneapolis, MN, 55455, by phone at 612-625-0447, or by email at mikem@me.umn.edu.

If you have any questions or concerns regarding this study and would like to talk to someone other than the researcher(s), **you are encouraged** to contact the University of Minnesota's Research Subjects' Advocate Line, D528 Mayo, 420 Delaware St. Southeast, Minneapolis, Minnesota 55455; (612) 625-1650.

You will be given a copy of this information to keep for your records.

Statement of Consent: I have read the above information. I have asked questions and have received answers. I consent to participate in the study. I give permission for the researchers to review my Department of Vehicle Services (DVS) records by providing my *MN Drivers License #*.

Signature:_____ Date: _____

MN Drivers License #: _____

Signature of Investigator:_____ Date: _____

APPENDIX K. RANDOM GAP STUDY INTRODUCTION

Our purpose is to investigate issues related to the use of active (or dynamic) traffic signs at rural intersections. Recent advances in technology have allowed the development of a system that can be placed near a STOP sign at a rural intersection to show drivers the state of traffic approaching the intersection on the main road. These signs are "smart" signs. This means the information on the sign changes in real time depending on the current traffic conditions near the intersection. This system presents information that helps you, the driver, make decisions about when to cross or turn at the intersection based on current traffic conditions. The diagram below shows a typical rural intersection where a smaller road crosses a larger, multi-lane road with fast-moving traffic. The signs you will see today can be placed at or near the STOP sign to help a driver waiting at the STOP sign make a decision about when to enter the intersection.

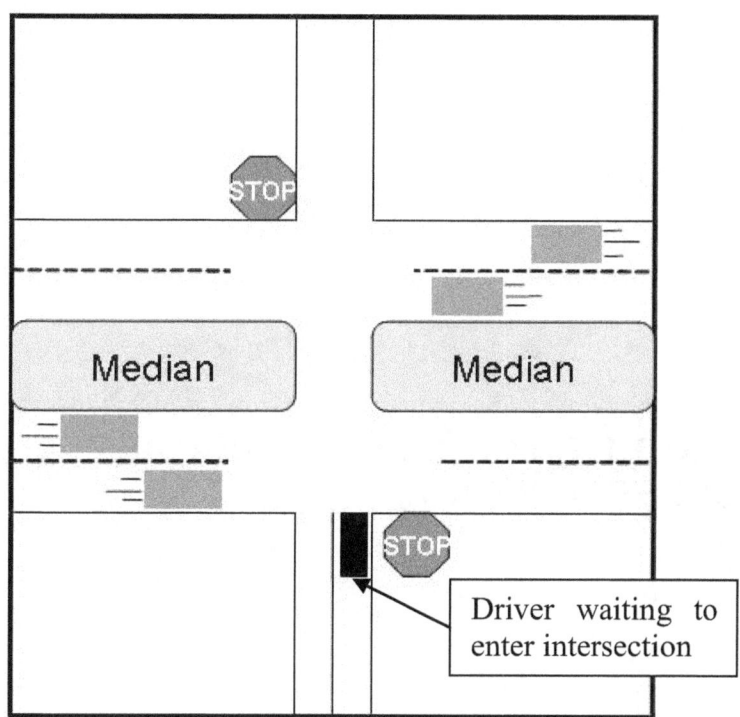

Please let the researcher know when you have finished reading this page.

APPENDIX L. MODIFIED COOPER-HARPER MENTAL WORKLOAD QUESTIONNAIRE

Think about the crossing maneuvers you just made in relation to the information provided by the "smart" sign present at the intersection.

1. Start in the bottom left-hand corner of the page and read the question. Follow the arrows depending on whether you answered "yes" or "no".
2. Continue to answer the questions until you arrive at the appropriate set of boxes on the right.
3. Choose a box on the right that best describes the level of effort and your perception of errors that may have occurred while trying to cross at the intersection using the "smart" sign. Mark an "X" under the number of the box.

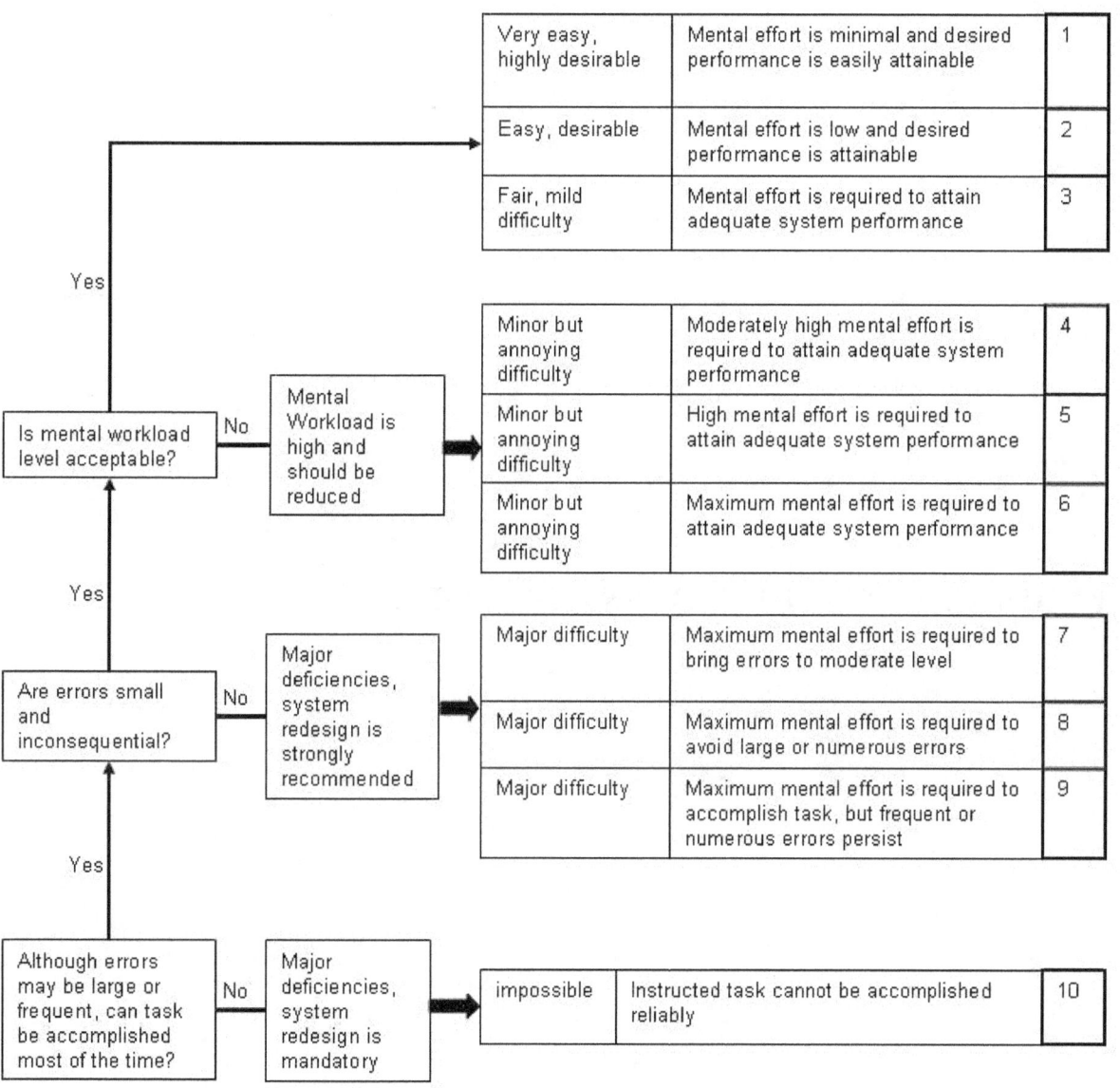

APPENDIX M. POST-DRIVE QUESTIONNAIRE

Please indicate how strongly you agree or disagree with the following statements. Answer these
questions in relation to the smart sign you just viewed at the intersection while driving.

1. I felt confident using this sign.

   ☐ Strongly Disagree   ☐ Disagree   ☐ Neutral   ☐ Agree   ☐ Strongly Agree

2. I felt it was confusing to use this sign.

   ☐ Strongly Disagree   ☐ Disagree   ☐ Neutral   ☐ Agree   ☐ Strongly Agree

3. Using this sign made me feel safer.

   ☐ Strongly Disagree   ☐ Disagree   ☐ Neutral   ☐ Agree   ☐ Strongly Agree

4. I trusted the information provided by the sign.

   ☐ Strongly Disagree   ☐ Disagree   ☐ Neutral   ☐ Agree   ☐ Strongly Agree

5. I like this sign.

   ☐ Strongly Disagree   ☐ Disagree   ☐ Neutral   ☐ Agree   ☐ Strongly Agree

6. The sign was reliable.

   ☐ Strongly Disagree   ☐ Disagree   ☐ Neutral   ☐ Agree   ☐ Strongly Agree

7. I felt this sign was easy to understand.

   ☐ Strongly Disagree   ☐ Disagree   ☐ Neutral   ☐ Agree   ☐ Strongly Agree

8. The sign's information was believable (credible).

   ☐ Strongly Disagree   ☐ Disagree   ☐ Neutral   ☐ Agree   ☐ Strongly Agree

9. This sign was useful.

☐	☐	☐	☐	☐
Strongly Disagree	Disagree	Neutral	Agree	Strongly Agree

10. I could complete the maneuver the same way without using the sign.

☐	☐	☐	☐	☐
Strongly Disagree	Disagree	Neutral	Agree	Strongly Agree

Continued on Next Page

11. Did you use the information on this sign to help you make your crossing decisions?

☐Yes ☐No

If **"yes"**, please explain what information you used or how you used the information to make your decision of when to cross?

If **"no"**, please explain why you did not use the information presented on the sign.

APPENDIX N. USABILITY QUESTIONNAIRE

Sign Description

You just viewed this sign at the intersection.

This sign shows an overview of the highway and the direction of travel of vehicles on the highway. This sign uses icons to indicate when traffic is detected near the intersection in each set of lanes (near and far lanes). When traffic is detected too close to the intersection in a set of lanes, a red block (indicating a vehicle) is lit up. At the same time, an icon indicates that it is unsafe to enter the intersection and which maneuvers might be dangerous. When a vehicle is detected approaching the intersection, but is not considered too close a yellow icon lights up (indicating the presence of a vehicle). This icon is yellow to indicate that it may be OK to cross, but that the driver should still proceed cautiously. If no vehicles are detected near the intersection, none of the icons are lit up. In this case, it may be ok to enter the intersection to cross over or turn right/left.

Sign with Different Messages	What Each Message Means
DIVIDED HIGHWAY	Do not enter the intersection; a vehicle is detected too close to the intersection in the near lanes (approaching from the left).
DIVIDED HIGHWAY	Do not enter the intersection; vehicles are detected too close to the intersection in both the near (approaching from left) and far lanes (approaching from right).
DIVIDED HIGHWAY	You may turn right; no vehicles detected approaching from the left in the near lanes. Vehicles are detected approaching from the right and are too close to the intersection; do not cross or turn left into the far lanes.
DIVIDED HIGHWAY	A vehicle is detected approaching from the left in the near lanes. You may be able to cross or turn, but proceed with caution.
DIVIDED HIGHWAY	No vehicles are detected approaching in the near (from the left) or far lanes (from the right). You may be able to cross or turn.

Please rate your opinion of the "smart" sign shown using all the items listed below.

Please refer to the "Sign Description" on the previous page if you need a reminder of how the sign works and the types of messages it presents. Remember that although multiple pictures are shown, this set of pictures represents only ONE sign that is capable of displaying several messages.

Example: If you thought the sign was very easy to use but required a lot of effort you might respond as follows:

Easy ☒ ☐ ☐ ☐ ☐ Difficult

Simple ☐ ☐ ☐ ☒ ☐ Confusing

Useful	☐ ☐ ☐ ☐ ☐	Useless
Pleasant	☐ ☐ ☐ ☐ ☐	Unpleasant
Bad	☐ ☐ ☐ ☐ ☐	Good
Nice	☐ ☐ ☐ ☐ ☐	Annoying
Effective	☐ ☐ ☐ ☐ ☐	Superfluous
Irritating	☐ ☐ ☐ ☐ ☐	Likeable
Assisting	☐ ☐ ☐ ☐ ☐	Worthless
Undesirable	☐ ☐ ☐ ☐ ☐	Desirable
Raising Alertness	☐ ☐ ☐ ☐ ☐	Sleep-inducing

Please let the researcher know you have finished this section.

APPENDIX O. RANKING QUESTIONNAIRE

Please rank the signs from 1 to 3. A rank of "**1**" indicates the sign is **most preferred** based on both your personal preference for it and on your assessment of how helpful you feel that sign is for making crossing decisions. A rank of "**3**" indicates the sign is **least preferred** based on your personal preference and your assessment of how helpful the sign is for making crossing decisions. Please refer to the information sheet that describes the meaning of the signs if you need a reminder of how each sign works and the types of messages it presents. Remember that, although multiple pictures are shown, each set of pictures represents only ONE sign that is capable of displaying several messages.

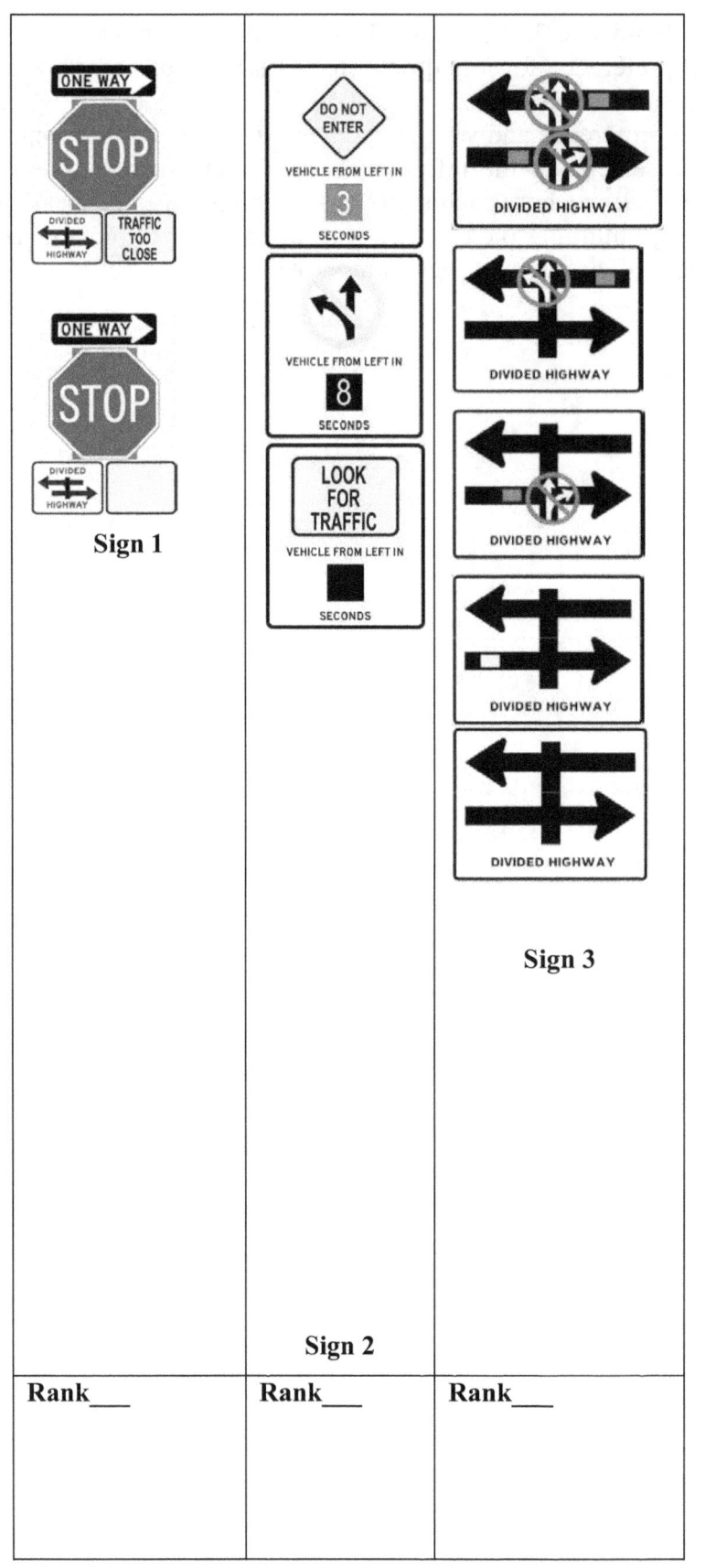

Sign 1

Sign 2

Sign 3

Rank___ Rank___ Rank___

APPENDIX P. RATING SCALE MENTAL EFFORT

Rating Scale Mental Effort

Please indicate, by marking the vertical axis below, how much effort it took for you to complete the task you've just finished

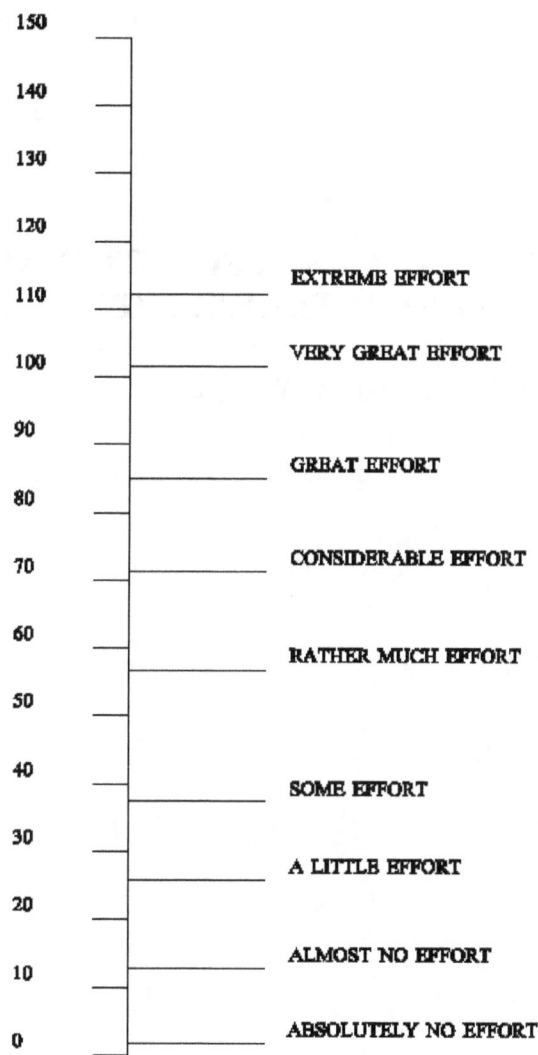

APPENDIX Q. SUMMARY OF CICAS-SSA FUNCTIONAL SCOPE AND DRIVER INFRASTRUCTURE INTERFACE (DII) TEST PROPOSAL

(by Janet Creaser and Nic Ward, 2007)

Table of Contents

1 **INTRODUCTION** **3**

2 **INFORMATION CONTENT** **4**

3 **TEST DESIGN BOUNDARIES** **6**

4 **BASIC MESSAGE SET (BMS)** **7**

5 **PROPOSED DII TEST DESIGN** **10**
   5.1   HAZARD SIGN 10
       5.1.1   Primary Information Content (see Table 1): 10
       5.1.2   Description of Sign Function 10
       5.1.3   Location of Sign at Intersection 11
       5.1.4   Limitations/Caveats 11
   5.2   COUNTDOWN SIGN 12
       5.2.1   Primary Information Content (see Table 1): 12
       5.2.2   Description of Sign Function 12
       5.2.3   Possible Locations of Signs at Intersection 17
       5.2.4   Limitations/Caveats 17
   5.3   ICON SIGN 18
       5.3.1   Primary Information Content (see Table 1): 18
       5.3.2   Description of Sign Function 18
       5.3.3   Location of Sign at Intersection 19
       5.3.4   Limitations/Caveats 19

6 **ADDITIONAL TEST DESIGN ISSUES** **20**

Introduction

Minnesota's Interregional Corridors (IRCs) are characterized by high-speed, high-volume roads that connect regional businesses, manufacturing, and tourist centers with rural districts. Intersection collisions are often fatal due to the high speeds, high volumes, and the heavy vehicles and trucks that routinely travel these routes (Donath & Shankwitz, 2001), making rural IRCs an important focus area for interventions. Rural IRC intersections in Minnesota can be either signal or sign controlled. The latter type is the focus of the Cooperative Intersection Collision Avoidance System – Stop Sign Assistance (CICAS-SSA). This is because a majority of rural intersection crashes occur at stop-controlled intersections. For example, of the 425 fatal rural crashes that occurred in Minnesota during 2002, 22% were at stop-controlled intersections compared to only 4.5% for signalized locations (Preston & Storm, 2003b). These data suggest rural stop-controlled intersections are an important safety consideration for drivers.

The goal of CICAS-SSA is to design and evaluate an infrastructure-based decision support system to help drivers safely negotiate rural stop-controlled intersections. Note that the system does not replace the gap decision process; instead, the driver remains responsible for choosing a safe gap and safely crossing the intersection. This system supports the minor-road driver in accepting safe gaps in mainline traffic at the intersection.

To minimize complexity, gaps are calculated from the minor-road driver to the next closest vehicle on the major road (Figure 30). In other words, it is assumed the first available "gap" is the one that is of greatest interest to the minor-road driver. Thus, the system actually monitors and provides driver support <u>in terms of the lag</u> defined by the headway of the lead vehicle in the mainline traffic with respect to the center of the minor vehicle lane (Figure 3). Consequently, gaps between leading and trailing vehicles farther down in the major-road traffic stream are not directly referred to in any of the interface solutions.

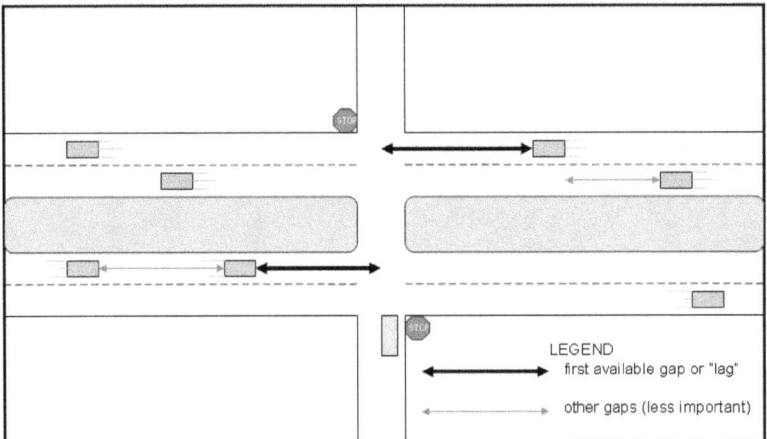

Figure 30. Stop-controlled intersection showing the first available "lag" and other gaps between leading and trailing vehicles.

The form of support is based on displaying information to the driver using changeable message signs (CMS) installed at the intersection on the minor road (and median). This information will relate to an estimation of the unsafe lag threshold (ULT) based on relevant parameters associated with each particular crossing minor-road driver. These relevant parameters will constitute a basic message set (BMS) that will be cooperatively transmitted between the approaching minor-road vehicles and the CICAS-SSA system that generates the support information to be presented within the CMS that functions as the driver-infrastructure interface (DII) for the system.

This memo presents an overview of the information content that is the basis of the DII concept and proposes several exemplars of DII test designs that are recommended for further evaluation in CICAS-SSA. The goal of this memo is to (1) solicit state and federal requirements for design modifications to achieve DII test formats that are consistent with MUTCD guidelines, and (2) obtain state and federal approval for the proposed system functionality and DII format in order to freeze the test design specification for the next stage of system evaluation.

6 Information Content

In order for the CICAS-SSA to be useful to drivers, the design of the DII information content must support the natural process of the gap acceptance task engaged by drivers at this class of intersection. As shown in Table 29, previous research in the Intersection Decision Support (IDS) project analyzed the primary information stages involved in negotiating a stop-controlled intersection (Laberge et al., 2003).

Table 29. Summary of Information Processing Stages for Intersection Negotiation Task.

A. **Alert minor-road and major-road drivers to the presence of the intersection.** Installing some of the interface solutions at intersections may have the secondary benefit of informing major-road and minor-road drivers that they are approaching an intersection. This could increase the probability that minor-road drivers stop before proceeding.

B. **Presence of vehicles that make up gaps.** Before a gap can be perceived and judged, the vehicle that forms the gap must be detected. Therefore, interface solutions can draw attention to and make major-road vehicles more salient.

C. **Convey speed, distance, and arrival time.** After a vehicle is detected, the temporal and spatial characteristics of the vehicle must be perceived. This is thought to occur based on estimates of speed, distance, and/or arrival time and forms the basis for perceiving gap size (Laberge et al., 2003). Interface solutions can make speed, distance, and arrival time of major-road vehicles explicit and improve the accuracy of gap perception for minor-road drivers.

D. **Size of available gaps.** Rather than force the minor-road driver to interpret the speed, distance, and/or arrival time information of individual major-road vehicles, interface solutions can convey gap size directly. This could take the form of time (s) or distance (ft) units.

E. **Judge whether a gap is safe.** Based on a finite list of factors (number of lanes, minor-road vehicle size, surface conditions) that influence the size of the safe gap, each major-road gap can be judged as safe or unsafe by the system. Interface solutions can inform the minor-road driver when a gap is unsafe. This takes the decision-making component away from the minor-road driver.

F. **Localize a safe gap and/or inform when a safe gap is about to arrive.** After the available major-road gaps have been identified as safe or unsafe, interface solutions can inform the minor-road driver which gaps in the traffic stream are safe and potentially when the next safe gap will arrive.

These information stages are presumed to be engaged by drivers in a logical sequence as shown in Figure 31. Accordingly, the temporal sequence of information acquisition required to support the intersection negotiation task is presumed to be hierarchical. For example, drivers must be aware that an intersection exists in order to trigger them to search for the presence of vehicles approaching that intersection. Similarly, the decision that a gap is safe is predicated on detecting a gap by observing parameters related to the approaching traffic, such as speed, distance from intersection or size of approaching vehicle.

Note that an intersection crash may result from a mistake at any stage in this information hierarchy. Indeed, it is not possible to predict at which stage a mistake may occur. Thus, it is reasonable to try to support higher information levels with the CICAS-SSA. This goal is achieved by presenting information that supports driver perception of vehicle parameters, gap size, and safe gap thresholds within the DII. Presenting information at higher levels must then logically support the lower-level information stages (detection of vehicle and intersection) upon which it is based (see Arrow in Figure 31).

Information Hierarchy:

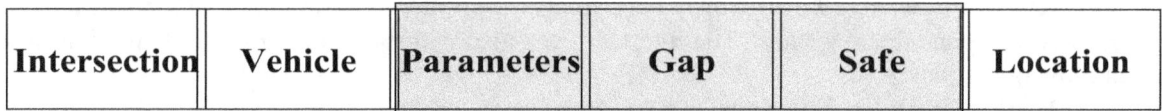

Figure 31. Information hierarchy supporting intersection negotiation task

3 TEST DESIGN BOUNDARIES

In order to set the foundation of the design for the CICAS-SSA system, it was necessary to establish boundaries for the scope and implementation of the system. The following is a list of presumed requirements and restrictions with respect to the system design and operations. These requirements will also serve as the foundation for developing a complete set of functional requirements for the CICAS-SSA system:

1. **Technically feasible**. Each interface solution is both technically possible and feasible. In other words, the technology exists and is readily available. Experimental technology is not considered and, when possible, the solutions assume or estimate parameters (i.e., driver age and PRT) to avoid adding additional technology that would measure them directly.

2. **Infrastructure solution**. For the reasons outlined in the original project proposal, the interface solutions are based only in the infrastructure (see Donath & Shankwitz, 2001).

3. **Applies to all stop-controlled intersections**. The interface solutions apply to all rural stop-controlled intersections regardless of specific geometry. This is important because designing for specific intersections would reduce the extent to which the solutions generalize to other intersections and other states.

4. **Minor-road driver is focus**. The interface solutions are targeted toward the minor-road driver and it is assumed the minor-road driver is responsible for any crash that would result. Improving the decision making of minor-road drivers will therefore reduce the likelihood that the same type of crash occurs.
5. **Minor-road driver stops**. The interface solutions assume the minor-road driver stops before proceeding. The solutions do not address situations where minor-road drivers violate the control device (willfully or due to inattention). A secondary benefit of some solutions is that the interface could increase the conspicuity of the intersection and reduce the likelihood of unintended sign violations.
6. **System does not impede traffic on the major road**. According to AASHTO (2001) guidelines for intersection sight distance, safe gaps are based on the assumption that major-road vehicles decelerate up to 30% to avoid a collision. The interface solutions assume the same limits by intending not to reduce the speed of major-road traffic by more than 30%.
7. **Minimal training required.** When possible, the interface solutions use stereotypic coding of information (color, frequency, symbols) to ensure meaning and that required actions are intuitive. This should increase understanding by drivers with minimal exposure and it is hoped this will reduce the need for training.
8. **Minimal additional signage required.** For some of the interface solutions, signage is needed to explain how the system works. To minimize cognitive overload, the number and complexity of the signs will be limited.
9. **Robust to winter conditions.** Each interface solution is visible in winter conditions, such as blowing and drifting snow. The interfaces can also withstand plowing and do not interfere with plow operations.
10. **Visible at night.** The interface solutions are visible at night.
11. **No interference with existing control devices**. The interface solutions complement and do not interfere with existing control devices. Obscuring a stop sign by placing a display in front of it would be an example of interfering with an existing control device.
12. **Use a prohibitive frame.** All the interface solutions use a prohibitive framework (e.g., "Do not turn left or cross"). This is important because a permissive framework is more liable if compliance leads to a crash (see Donath & Shankwitz, 2001).
13. **Median monitored for occupancy.** The system can support only one minor road vehicle a time. This requires that the system has a method of identifying which minor road driver is the relevant target. For example, this may require specifying a fixed area at the stop signs that demarcate the assume location and starting point of the stopped target vehicle. This in necessary to have a fixed basis for calculating movement time in the ULT calculations. In addition, the system will need to (1) monitor and alternate access for cross traffic from the minor road and (2) monitor the occupancy state of the median and prohibit entry if the median is occupied.
14. **Accommodate major-road traffic that exits at the minor road.** To the extent that cases of mainline traffic exiting from the minor road intersection is considered a relevant or hazardous condition, it will be necessary to identify cases of mainline traffic entering the intersection to cross like minor road vehicles. For such cases, it will be necessary to monitor turning pockets from the main road into the minor road intersection.

7 Basic Message Set (BMS)

The fundamental component of CICAS-SSA is the cooperation between the infrastructure-based system and the approaching minor road vehicles. Specifically, in contrast to the IDS project which used a static unsafe lag threshold (ULT) as a basis for advising drivers, the CICAS-SSA project uses dynamic thresholds that are individualized to the specific parameters of each minor-road driver. Thus, the CICAS-SSA project includes several tasks that support the definition of a basic message set (BMS) that comprises the most relevant, feasible, and parsimonious set of parameters to predict unsafe lag thresholds for a broad class of minor road drivers.

Parameters for the BMS will be based on data available from sensors integrated at the intersection that communicate information about mainline traffic (speed, distance, gaps, and lags), median occupancy, and classification of the minor road target vehicle. Additional information may be available from vehicle (or key) tag information regarding the type of driver and expected crossing maneuver. The resulting BMS can be communicated with DSRC to the central processor for the system to calculate the ULT for a target case.

It is important to note that the constitution of the BMS will depend on the range of sensors and data that can be transmitted to the CICAS-SSA. In this regard, three scenarios can be envisaged each with a different operating BMS:

A. *Full BMS and complete ULT algorithm – Minor Road Traffic.* Minor road vehicles equipped with all system components necessary for cooperative communication with the CICAS-SSA (e.g., DSRC, tag). In this case, all parameters can be computed for the CICAS-SSA to function with the complete ULT algorithm.

B. *Full BMS and complete ULT algorithm – Mainline Traffic.* Although the mainline will not be equipped with sensors to classify vehicle type, the full BMS and complete ULT algorithm is still applicable to those mainline vehicles entering the median in order to execute a left turn (LTAP/OD) that are equipped with DSRC and tags to communicate vehicle type (and location) directly to the CICAS-SSA.

C. *Partial BMS and reduced ULT algorithm – Minor Road Traffic.* Whereas vehicle classification is available for all minor road vehicles because the classification sensors are embedded along the minor road, the CICAS-SSA will respond to minor road vehicles that do not have cooperative communication components with a reduced ULT algorithm (excluding those parameters that would otherwise be communicated from the tag).

Although the specific elements of the BMS cannot be specified at this stage of the project, this table provides a description of several parameters that the literature indicates are related to safe gap thresholds.

For those solutions that rely on the safe gap, the generic description is:

$$t_G = t_{PRT} + t_{MT} \quad \text{(Laberge et al., 2003)}$$

Where

t_G = the safe "lag" (s)

t_{PRT} = the perception response time (PRT) needed for the minor-road driver to detect, perceive, and accept a gap and initiate the maneuver (s)

t_{MT} = the time required to accelerate to speed and cross the distance needed to clear or enter the major road (s)

30. Summary of Potential Parameters Related to Safe Gap Thresholds.

- Age = older drivers may be slower to detect, perceive, and accept a gap as well as slower to accelerate and complete their maneuvers (Keskinen et al., 1998; Lerner et al., 1995; Olson, 2002; Wagner, 1965). This will result in increased t_{PRT} and t_{MT}. It is assumed there is no way to detect driver age, so older drivers will be the default case. This is justified because older drivers have been identified as the highest-risk group for whom the system is being designed (Laberge et al., 2003). The exact adjustment to the size of the safe gap for older drivers will be determined during the detailed design stage.

- Distraction = distracted drivers could be slower to detect gaps and decide which is acceptable and may not consider all factors when making decisions. This may increase t_{PRT}. It is assumed there is no way to detect driver distraction, and therefore it is not a parameter that is used when calculating the safe gap.

- Maneuver type = left turns are a more complicated decision and require a greater distance to cross as well as more time to merge with far-side traffic. This will increase t_{PRT} and t_{MT}. It is assumed there is no way to detect maneuver type for the minor-road vehicle in advance of the intersection. Therefore, the default maneuver type will be the left turn. According to AASHTO (2001), left-turn maneuvers require a safe gap that is 1 s longer than crossing or right-turn maneuvers and a greater adjustment is needed for larger vehicles.

- Lanes = more lanes to cross for left-turn and crossing maneuvers will increase t_{MT}. The number and width of the lanes at each site can be calculated in advance and adjustments made to the safe gap. According to AASHTO (2001), 0.5 s is required for each 12-ft lane the driver is required to cross and a greater adjustment is needed for larger vehicles

- Minor-road vehicle size = larger vehicles are slower to accelerate and will require a larger gap to enter or cross, increasing t_{MT}. If minor-road vehicle size can be detected from the infrastructure, some interface solutions can accommodate safe gaps that increase as vehicle size increases. According to AASHTO, the safe gap for passenger cars is 7.5 s, 9.5 s for single-unit trucks, and 11.5 s for combination trucks

- Surface conditions = slippery roads will reduce traction and increase t_{MT}. If the infrastructure can detect pavement surface conditions at the intersection, some interface solutions can accommodate safe gaps that increase as surface conditions become more slippery.

Proposed DII Test Design

The IDS project proposed a number of CICAS-SSA concepts in terms of information displays. These concepts were intended to represent the types of information that could be presented to drivers as part of a CICAS-SSA system (see Table 29). These concepts were never intended to represent final design elements. Instead, the CICAS-SSA project is expected to consult with state and federal officials to decide on the design elements for the DII while retaining the design benefits of the original concepts. Toward that end, UMN is proposing to evaluate several DII concepts are discussed below. Each concept is described in detail including a (1) description of the information content provided (see Table 29), (2) summary of the system function linked to this DII, (3) diagram of the DII configuration with respect to the intersection, (4) optional variations in design for the DII and system, and (5) some suspected limitations and caveats for the DII and system.

7.1 Hazard Sign
7.1.1 Primary Information Content (see Table 29):

    A. Alert minor-road drivers to the presence of the intersection.

    B. Presence of vehicles that make up lag.

    E. Judge whether a lag is safe.

7.1.2 Description of Sign Function

The Hazard sign concept consists of a changeable message sign located below the stop sign (see Table 3, Figure 3A). The sign flashes the message "Traffic Too Close" on a yellow background when the lead major-road vehicle in either direction is within the arrival time that defines the safe lag threshold for the near and far lanes. The message flashes until the hazardous condition(s) has passed and the illuminated period of the flash cycle is 500 ms. If the system does not detect traffic within the lag thresholds of the near and far side lanes no message is displayed and the background is yellow. This design supports the detection of major-road vehicles approaching the intersection in either the near lanes or the far lanes or both, but does not provide specific lag information related to perception of lag size or assessment of lag safety.

It is possible that this sign could be used both at the stop sign, as originally intended, and also in the median where it could be attached to the Yield sign and provide warnings for drivers who make two-stage maneuvers (see Figure 32B). If it were also used in the median, the median sign would only flash when far-side vehicles were inside the lag threshold.

Table 31. Hazard sign showing "on" and "off" functions.

	Condition	
	Driver At STOP Sign	**Driver In Median**
Near-side vehicle <u>inside</u> lag threshold lag AND far-side vehicle absent, or inside, or outside lag threshold -OR- Far-side vehicle <u>inside</u> lag threshold AND near-side vehicle absent or outside lag threshold	ONE WAY → STOP / DIVIDED HIGHWAY / TRAFFIC TOO CLOSE	YIELD / TRAFFIC TOO CLOSE
Near-side vehicle <u>outside</u> lag threshold AND far-side vehicle <u>outside</u> lag threshold	ONE WAY → STOP / DIVIDED HIGHWAY	YIELD

7.1.3 Location of Sign at Intersection

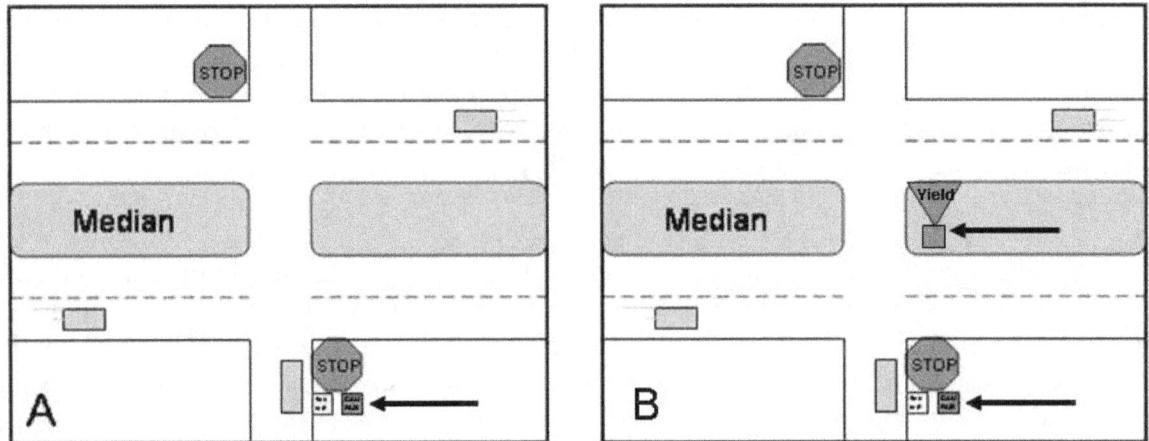

Figure 31. Possible locations of Hazard sign at intersection. A) Attached to Stop Sign post only. B) Located on Stop sign Post and in median on Yield sign post.

10

7.1.4 T Limitations/Caveats

- The sign will only stop flashing when vehicles are outside the safe gaps for both sets of lanes; heavy traffic in both directions or a single direction will result in the sign continuing to flash "Traffic Too Close."
- The sign does not provide specific information to help minor-road drivers make better gap acceptance decisions.
- Reading text and interpreting symbols can be cognitively demanding and error prone.
- There may be some issues related to text legibility and comprehension (especially for older drivers).
- When traffic volumes are high, there exists a chance that no safe lags will be present and the DO NOT ENTER symbol is shown for long periods of time. This could result in a long queue on the minor road and drivers could ignore the display as both wait time and driver frustration increase.

7.2 Countdown Sign
7.2.1 Primary Information Content (see Table 29):

A. Alert minor-road drivers to the presence of the intersection.

B. Presence of vehicles that make up lags.

C. Conveys arrival time.

E. Judge whether a lag is safe.

7.2.2 Description of Sign Function

The Countdown sign uses icons to indicate the safety of an available lag and a countdown timer that shows the arrival time of major-road vehicles approaching the intersection (see Table 4). The association between arrival time and the prohibitive message should help minor-road drivers learn when it is unsafe to complete each maneuver. Two displays are used for this concept and there are two options for locating this sign at the intersection. First, one sign can be placed on the left side of the intersection when the driver is at the stop sign and the median sign can be placed on the driver's right side when they are in the median (see Figure 4A). This configuration allows drivers to monitor the signs and the traffic simultaneously. Alternatively, the sign providing information when the driver is at the stop sign can be placed in the left median across the near lanes and the sign providing information when the driver is in the median can be placed on the far side of the intersection (see Figure 4B). This second configuration prevents the displays from blocking the sign lines to the approaching traffic yet still allows them to be viewed while monitoring traffic in the appropriate directions.

The near-side sign information relates to vehicles that are tracked in both the near and far lanes. The far-side sign only tracks vehicles in the far lanes and is only used once the driver is in the median. If it is unsafe to cross the near lanes while a driver is at the stop sign, the median sign could convey the same information as the near-side sign so as not to confuse the driver about the correct course of action while at the stop sign (i.e., even if it is considered safe to cross the far lanes a driver cannot do so if it is unsafe to cross the near lanes). The near-side sign includes the words "Vehicle from left in" while the far-side sign includes "Vehicle from right in" above the timer countdowns.

The time-to-arrival for the major road vehicle is counted down in 1 s increments. When a detected vehicle is outside the lag threshold, the timer background is black and shows the time in white lettering. Because the sensor system will only track vehicles out to a certain point on the major road, the timer may not always show a time. If no vehicle is detected within the sensors' range, the background will be black and show no time at all, indicating no vehicles are detected near the intersection. When a detected vehicle is inside the lag threshold, the timer background is red and shows the time in white lettering.

A different icon is presented depending on where traffic is located relative to the intersection for each set of lanes, and depending on where the driver is (i.e., at stop sign or in median). Yellow is used as the warning color for this sign and the icon shapes could be alternated for each icon message to make them more quickly distinguishable from each other.

7.2.2.1 Driver at Stop Sign

When the driver is stopped at the stop sign, three icons are used to indicate the state of traffic at the intersection. First, when a vehicle is inside the lag threshold of the near lanes (or inside the safe threshold for both the near and far lanes), the timer background is red and an icon that represents it is not safe to enter the intersection is presented. There are three possible options for presenting this message (see Table 32; Options 1-3).

Second, when a vehicle is inside the far-lanes lag threshold but no vehicle is inside the near-lanes lag threshold a yellow circle and slash over an icon showing a left-turn and straight arrow is presented. This means it is not safe to cross over the intersection or turn left. The timer background for the near sign will be black. It is implied that it may be safe to turn right or cross to the median. There are two possible options for this icon (see Table 32; Options 4-5).

Finally, when no vehicles are tracked in either the near or far lanes (or tracked vehicles are outside the lag threshold), a yellow rectangle or yellow diamond can be presented with the words "LOOK FOR TRAFFIC" inside (see Table 32; Options 6-7). The timer background will be black.

7.2.2.2 Driver in Median

Only two messages are presented when the driver is in the median (see Table 33). First, when a vehicle is within the far lanes safe lag threshold, an icon indicating it is not safe to enter and that the driver should wait is presented and the timer background will be red (Options 8-10). Second, when no vehicle is tracked in the far lanes or no vehicle is within the lag threshold, the yellow rectangle or yellow diamond with the words "LOOK FOR TRAFFIC" can be presented in conjunction with a black timer background.

The Countdown sign supports the detection of vehicles and lags, provides an absolute location of approaching vehicles by presenting the actual time for a vehicle to reach the intersection, allows the driver to monitor changes in the arrival time and, thus, the lag size, and assesses the safety of available lags.

Table 32. Countdown sign showing icon options for each traffic message when driver is at the STOP sign.

Location of Traffic when Driver is at STOP sign	Icon Options for each Message*		
	Option 1	Option 2	Option 3
Near-side vehicle <u>inside</u> lag threshold AND far-side vehicle absent, outside, inside lag threshold	This icon uses the yellow warning diamond and tells the driver not to enter the intersection when the threshold is small.	This icon shows a crash, indicating to the driver that entering the intersection with a small lag threshold could possibly lead to a crash.	This icon shows a hand and says "wait", indicating to the driver that they should wait until the threshold changes.

	Option 4	Option 5	
Near-side vehicle <u>outside</u> lag threshold AND far-side vehicle <u>inside</u> lag threshold	In this option, the "do not turn left or go straight" arrows and the prohibitive circle and slash are black. They are presented inside the yellow warning diamond to warn about completing these maneuvers when visible.	In this option the arrows are black and the prohibitive slash and circle are yellow, warning the driver about completing these maneuvers when visible.	
	Option 6	**Option 7**	
Near-side vehicle <u>outside</u> lag threshold AND far-side vehicle <u>outside</u> lag threshold	In this option, the "look for traffic" message is placed inside a yellow rectangle, which indicates the driver should use caution when crossing.	In this option, the "Look for traffic" message is placed inside the yellow warning diamond, which indicates the driver should use caution when crossing.	

*Although any combination of these icon options could be used, it is likely that the choice for each message would use a different icon shape to make each distinguishable from the others (e.g., diamond for "do not enter" message, circle for "do not turn left or cross" message, and the rectangle for the "look for traffic" message).

Table 33. Countdown sign showing icon options for each traffic message when driver is in the median.

Location of Traffic when Driver is in Median	Possible Icon Options*		
	Option 8	Option 9	Option 10
Far-side vehicle inside lag threshold	[icon: truck crash symbol, VEHICLE FROM RIGHT IN 3 SECONDS]	[icon: DO NOT ENTER, VEHICLE FROM RIGHT IN 3 SECONDS]	[icon: WAIT hand, VEHICLE FROM RIGHT IN 3 SECONDS]
	Option 11	Option 12	
Far-side vehicle <u>outside</u> lag threshold	[icon: LOOK FOR TRAFFIC, VEHICLE FROM RIGHT IN 11 SECONDS]	[icon: LOOK FOR TRAFFIC diamond, VEHICLE FROM RIGHT IN 11 SECONDS]	

*Final icon message set for near sign and median sign would be the same.

7.2.3 Possible Locations of Signs at Intersection

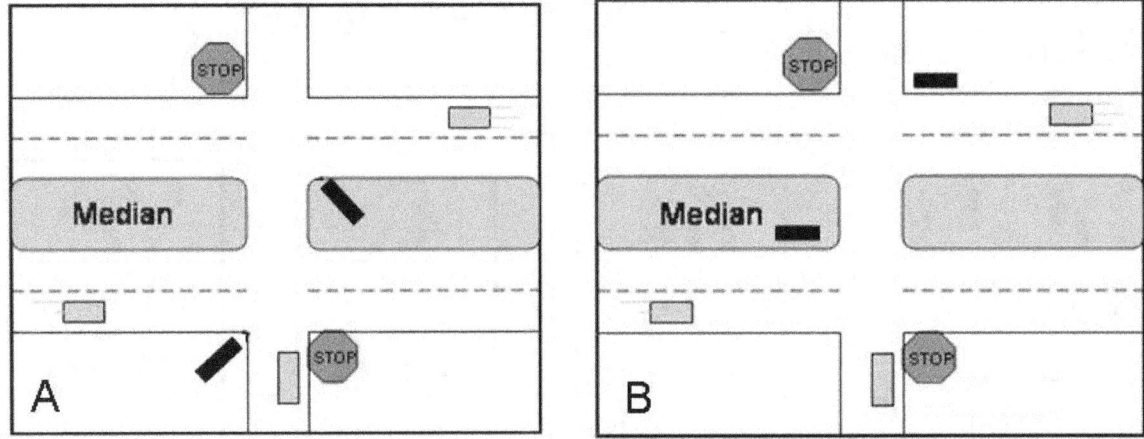

Figure 32. Location of Countdown sign at intersection. A) The near-side sign is placed to the driver's left when at the stop sign while the median sign is placed on the driver's right in the median. B) The signs are still placed to the left and right of the driver, but across the lanes.

7.2.4 Limitations/Caveats

- Drivers may not be able to interpret absolute values for arrival time. In other words, drivers may not perceive arrival time as a numerical value or may not be sensitive to small differences in arrival time.

- Reading text and interpreting symbols can be cognitively demanding and error prone.

- There may be some issues related to text legibility and comprehension (especially for older drivers).

- There exists a potential for rapid fluctuation between symbols/messages (especially when traffic volumes are high). This could be confusing for the driver.

- When traffic volumes are high, there exists a chance that no safe lags will be present and the DO NOT ENTER symbol is shown for long periods of time. This could result in a long queue on the minor road and drivers could ignore the display as both wait time and driver frustration increase.

7.3 Icon Sign
7.3.1 Primary Information Content (see Table 29):

    A. Alert minor-road drivers to the presence of the intersection.

    B. Presence of vehicles that make up lags.

    D. Size of available lags.

    E. Judge whether a lag is safe.

7.3.2 Description of Sign Function

The Icon sign consists of an iconic image of the intersection that warns drivers of approaching vehicles and their relative distance from the intersection (see Table 34). It also displays prohibitive icons indicating the maneuvers driver should not make when unsafe lags are detected. This sign provides warning information about approaching vehicles using a yellow, then a red icon as the detected vehicle approaches. The yellow icon is illuminated when a vehicle is detected near the intersection, but has not yet crossed the safe lag threshold. The path icons and prohibitive circle-slash are not shown in this case. Once the vehicle crosses the safe lag threshold, a red icon is illuminated and a red prohibitive circle and slash is displayed over a white path indicator showing directions of travel for the lanes in which the vehicle is detected. The near and far lanes are monitored independently and two signs are used, one at the stop sign and another in the median to assist with all stages of the crossing or turning at the intersection. There are two variations on where this sign can be located (see Figure 34). The bottom portion of the median sign is inactive and is faded by 60% when a driver is in the median, allowing them to focus on the appropriate portion of the sign.

Notes: It may not be possible to render the red and yellow icons with appropriate lamination and size to meet the visual needs of all drivers. In this case, alternative warning level presentations will be considered.

Table 34. Icon sign.

	Condition			
	Driver At STOP Sign		Driver In Median*	
Icon Sign	Near-side vehicle within safe lag AND far-side vehicle within safe lag	[icon]	Far-side vehicle within safe lag	[icon]
	Near-side vehicle approaching but outside safe lag AND far-side vehicle within safe lag	[icon]	Far-side vehicle approaching but outside safe lag	[icon]
	Near-side vehicle outside safe lag AND far-side vehicle outside safe lag	[icon]	Far-side vehicle outside safe lag	[icon]

7.3.3 Location of Sign at Intersection

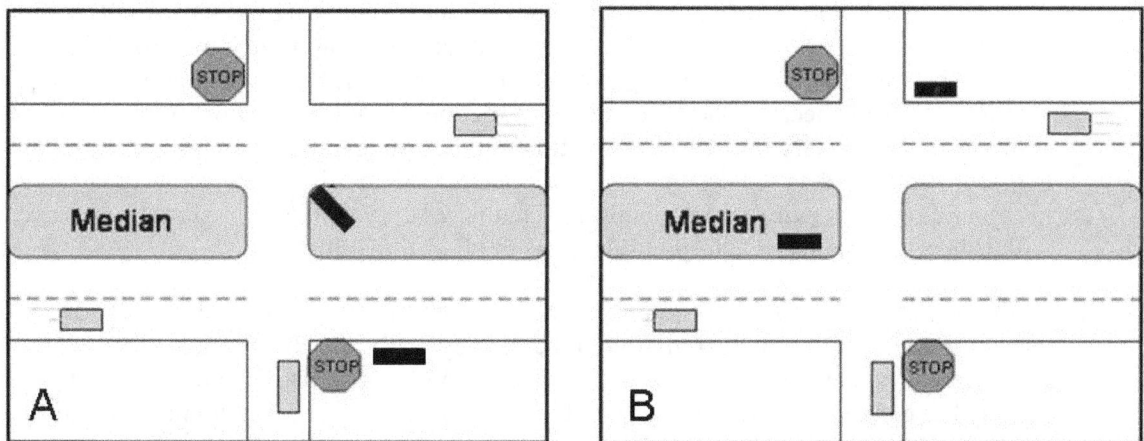

Figure 33. Location of Icon sign at intersection. A) The near-side sign is placed to the driver's right when at the stop sign while the median sign is placed on the driver's right in the median. B) The signs are placed to the left and right of the driver, but across the lanes.

7.3.4 Limitations/Caveats

- The sign may need to be large in order to convey the necessary information. Its size may block the minor-road driver's view of traffic.

- Interpreting symbols can be cognitively demanding and error prone when multiple levels of information are presented.

- There may be some issues related to icon legibility and comprehension (especially for older drivers).

- There exists a potential for rapid fluctuation between symbols (especially when traffic volumes are high). This may be confusing for the driver.

- When traffic volumes are high, there exists a chance that no safe lags will be present and the DO NOT ENTER symbol, for one or both sides of the intersection, is shown for long periods of time. This could result in a long queue on the minor road and drivers could ignore the display as both wait time and driver frustration increase.

8 Additional Test Design Issues

1. All interface solutions must communicate the current system state to minor-road drivers. This is important because it is possible the system could fail and the minor-road driver needs to be able to differentiate between "no signal" because the intersection is safe versus "no signal" because the system is not functioning. Although most of the interface solutions fail to specify how this information will be conveyed, it is an issue that will be considered during detailed design. Another issue related to this is how drivers may interpret the withdrawing of a prohibition. That is, when a "do not go" signal is removed, will drivers automatically interpret this as an implied signal to "go.

2. It is unclear how the display is activated and whether an activation region is needed so drivers know whether the information in the display applies to them or a lead vehicle. A related requirement is that the DII display only recommend gaps when the previous vehicle has cleared the intersection (i.e., they are not waiting at the median crossover).

3. When traffic volume is high, timing elements (countdown timers such as those in the Countdown) will fluctuate as the lead vehicle passes the intersection and the next lead vehicle is tracked. Figure 35A shows two near-side (coming from the left) vehicles approaching. The (original) Countdown concept would initially display the arrival time of vehicle A (9 s). After vehicle A enters the region that defines the safe gap (i.e., 8 s), the arrival time would be highlighted in red and the corresponding speed and image would be displayed. After vehicle A passes the intersection (Figure 35B), the Countdown concept would display the arrival time of vehicle B (as the next lead vehicle). If vehicle B is initially tracked at an arrival time greater than the safe gap, the prohibitive message would briefly change to the default CAUTION, but then quickly change back to DO NOT ENTER after vehicle B enters the safe gap. This rapid fluctuation would increase as traffic volumes increase and more lanes exist on the major road. It will be difficult for minor-road drivers to understand which vehicle is being tracked and comprehend the rapid changes in prohibitive messages. One solution may be to track a swarm or group of vehicles with the hope of minimizing rapid fluctuations.

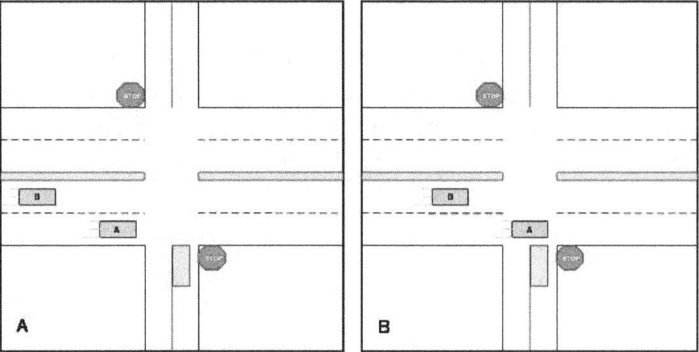

Figure 34. Vehicles approaching close together may result in rapid fluctuation as Vehicle A arrives at the intersection and Vehicle B begins to be tracked very shortly afterwards.

4. How do we treat major road traffic that turn into the intersection and then function as minor road traffic crossing through the median? At this stage, the mainline traffic cases are treated as minor road traffic (see below).

Mainline Traffic Entering Minor Road

LTAP/OD

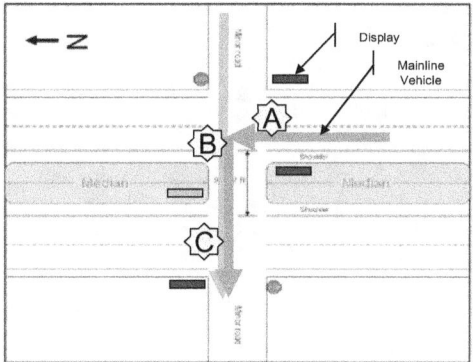

- The system displays the 'lag' for the mainline vehicle at point A.
- When this vehicle enters the left turn lane and moves into the median (Point B), it then assumes the same crossing manouvre as the minor road vehicle.
- While in the median, the mainline vehicle could utilize the median display.
- Therefore, the display algorithm must function for minor road vehicles <u>and</u> mainline traffic moving into the median to cross onto the minor road and avoid a LTAP/OD crash (Point C).
- This means that the operating system will need vehicle classification of mainline vehicles (in the left turn lanes) assuming that vehicle type becomes a part of the system algorithm (for the minor road vehicles).

22

U.S. Department of Transportation
ITS Joint Program Office-HOIT
1200 New Jersey Avenue, SE
Washington, DC 20590

Toll-Free "Help Line" 866-367-7487
www.its.dot.gov

U.S. Department of Transportation

Research and Innovative Technology Administration

www.ingramcontent.com/pod-product-compliance
Lightning Source LLC
Chambersburg PA
CBHW080241180526
45167CB00006B/2371